# 3rd International Conference on Accelerating Biopharmaceutical Development 2013



Coronado Island, California, USA
24-27 February 2013

ISBN: 978-1-5108-1598-8

**Printed from e-media with permission by:**

Curran Associates, Inc.
57 Morehouse Lane
Red Hook, NY 12571

**Some format issues inherent in the e-media version may also appear in this print version.**



Printed by Curran Associates, Inc. (2015)



AIChE
120 Wall Street, FL 23
New York, NY 10005-4020

Phone: (800) 242-4363
Fax: (203) 775-5177

www.aiche.org

**Additional copies of this publication are available from:**

Curran Associates, Inc.
57 Morehouse Lane
Red Hook, NY 12571 USA
Phone: 845-758-0400
Fax: 845-758-2634
Email: curran@proceedings.com
Web: www.proceedings.com

# TABLE OF CONTENTS

## QBD ROUNDTABLE



## KEYNOTE



## POSTER SESSION

**Optimizing the Speed and Throughput of Therapeutic Antibody Development**

Bailey, K., Regeneron

Originally presented on: 2/24/2013 16:45 - 17:45

Advances in target discovery and validation coupled with the ability to generate fully human antibodies against these targets is increasing the rate with which biotechnology companies can move promising new therapeutics into clinical development. To meet this challenge, new rapid development paradigms are needed to identify the best candidates and deliver quality therapeutics to patients. Regeneron's speed to clinic development strategy along with our approach for late-stage therapeutic development will be discussed in conjunction with an overview of the technologies used to facilitate these efforts.

# *Biotechnology for the Next Decade:*
## *Meeting Society's Expectation for Pharmaceutical Innovation*

Joseph P. Miletich, M.D., Ph.D.
Senior Vice President, Research and Development
Amgen Inc.

## Industry-Wide R&D Productivity Has Been Declining for Decades: A New Approach Is Needed

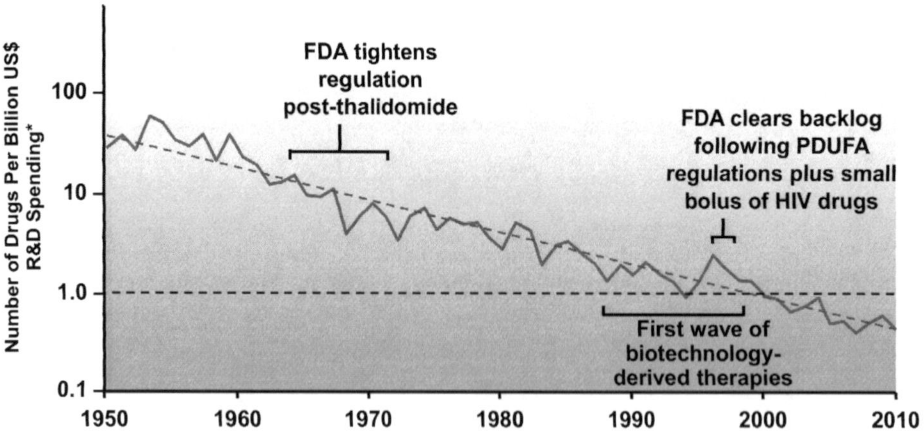


*Inflation adjusted
PDUFA = Prescription Drug User Fee Act
Source: JW Scannell, Diagnosing the decline in pharmaceutical R&D efficiency, *Nature Reviews Drug Discovery*, 2012, v11, pg191

## Biologics manufacturing is entering a period of Disruptive Innovation

New technologies typically launch as complex, hard to operate and user-dependent

Over time processes become defined/refined; methods become standardized

Size shrinks, reliability increases, the technology is disbursed

We are amidst a step-function change in our manufacturing capabilities

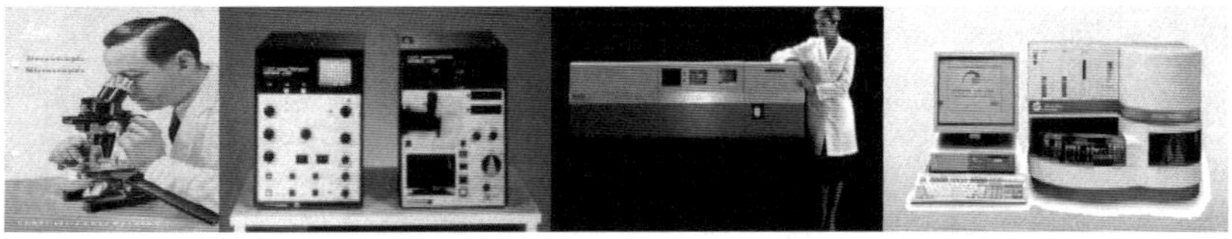

***Disruptive innovation is common with new technologies and has a predictable maturation cycle***

*If the mass of people hesitate to act, strike thou in swift with all boldness; the noble heart that understands and seizes quick hold of opportunity can achieve everything.*

*- Johann Wolfgang von Goethe*

## Disruptive technologies are changing our approach to biologic discovery and drug development

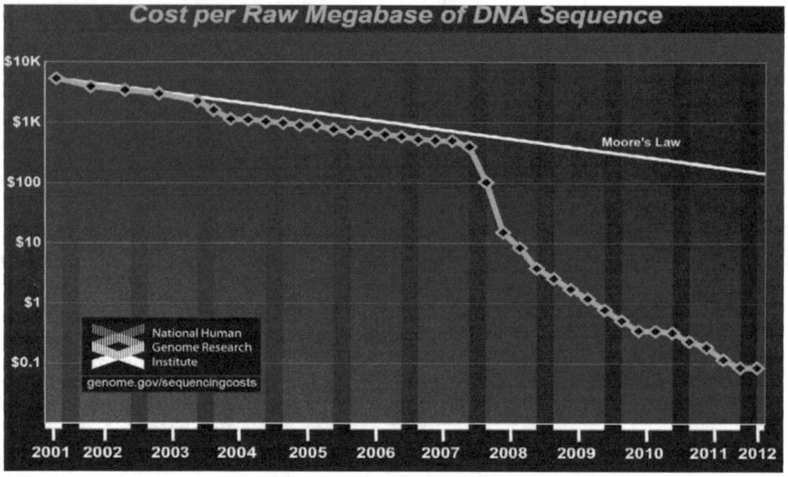


***The safe and ethical introduction of new therapeutics is our best mechanism for learning***

## *Driving Innovation In An Increasingly Challenging Environment*

**Challenges**

- Regulatory landscape
- Health economics
- Discovery research productivity

**Opportunities**

- Lower-cost, flexible manufacturing
- New modalities
- Improved clinical trial design

**Realities**

- Consolidation
- Biosimilars
- Globalization

***Is It Worthwhile Pursuing New, Innovative Medicines?***

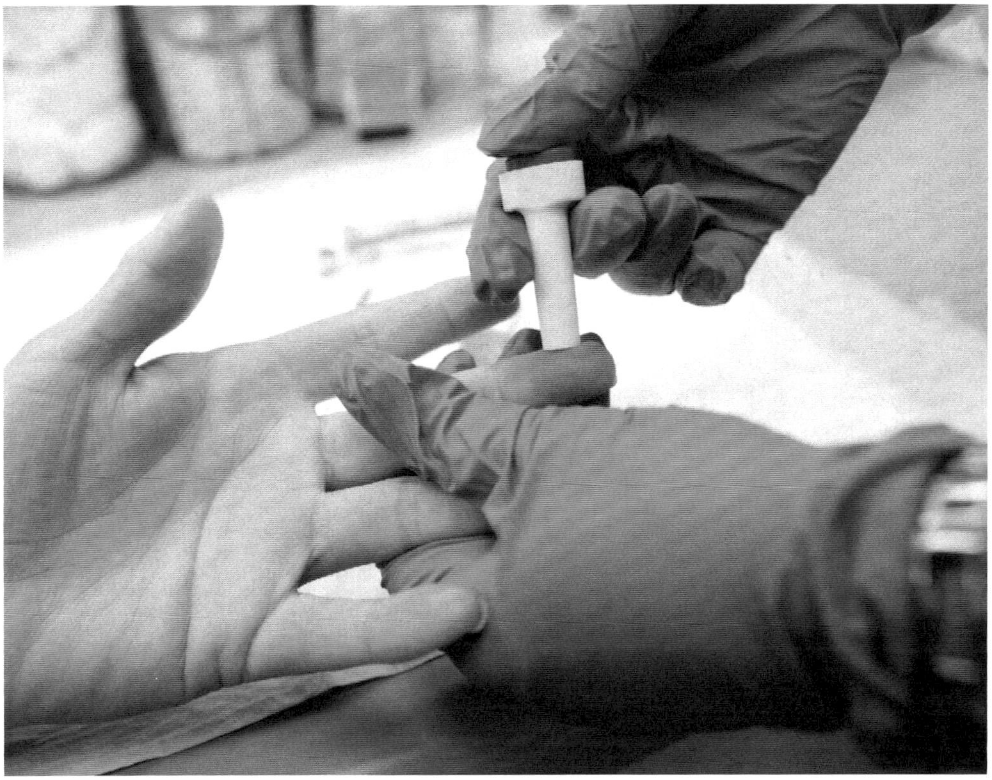

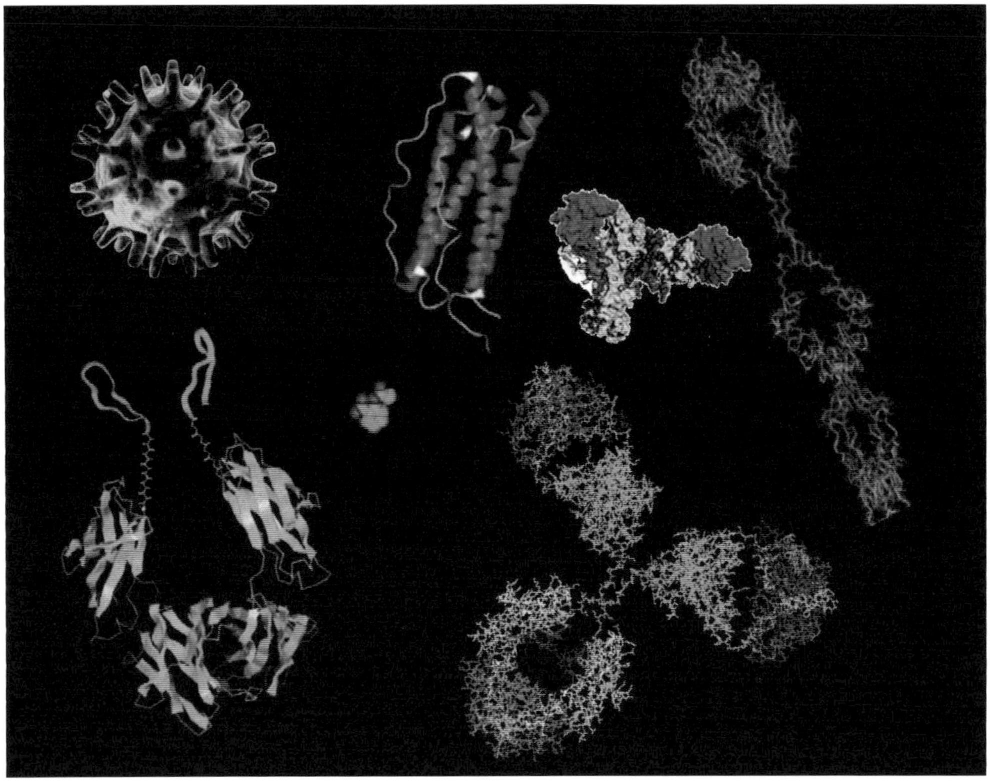

Diagnosability
Convertibility
Scalability
Customization
Modularity
Integrability
Reconfigurable
Manufacturing
Systems

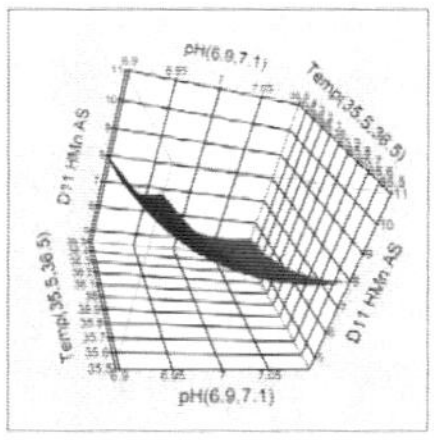

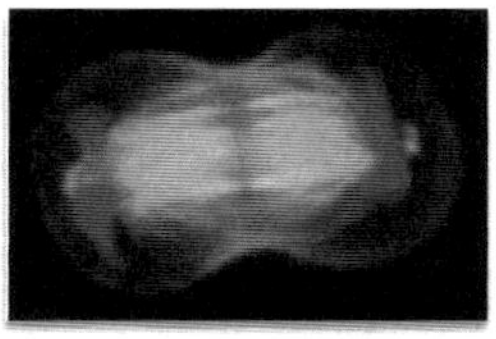
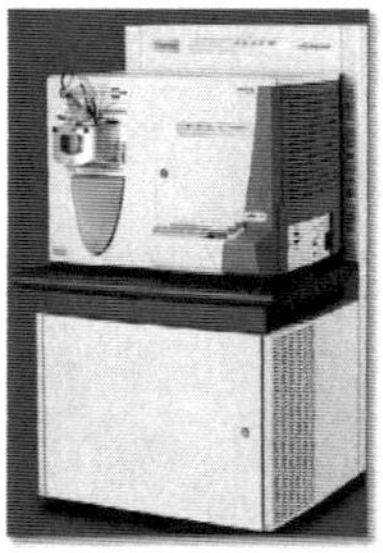

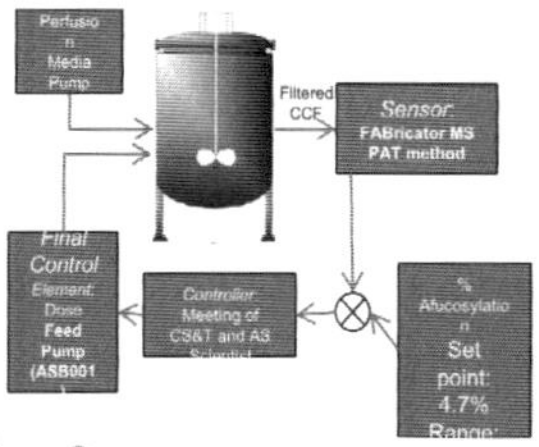

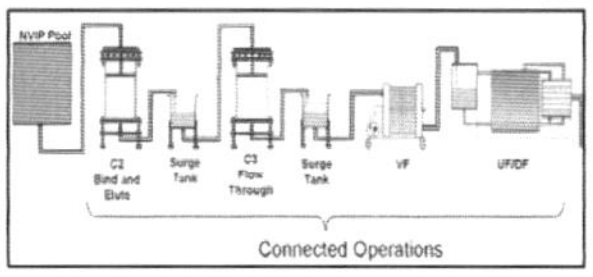

## *Is It Worthwhile Pursuing New, Innovative Medicines?*

ALZHEIMER'S DISEASE
ANEMIA ASSOCIATED WITH CHRONIC RENAL FAILURE
ANESTHESIA CARE SEDATION
ANGIOEDEMA ATTACKS IN ADOLESCENTS/ADULTS WITH HEREDITARY ANGIOEDEMA
ASTHMA
ATTENTION-DEFICIT/HYPERACTIVITY DISORDER
BENIGN PROSTATIC HYPERPLASIA
BONE METASTASES FROM SOLID TUMORS
BREAKTHROUGH CANCER PAIN
BREAST CANCER
CHEMOTHERAPY-INDUCED NAUSEA AND VOMITING
CHOREA DUE TO HUNTINGTON'S DISEASE
CHRONIC GOUT (HYPERURICEMIA)
CHRONIC MIGRAINE
CHRONIC OBSTRUCTIVE PULMONARY DISEASE
CROHN'S DISEASE
DYSLIPIDEMIA AND HYPERTRIGLYCERIDEMIA
DYSLIPIDEMIA, HYPERTRIGLYCERIDEMIA AND HYPERLIPIDEMIA
FIBROMYALGIA
GASTRIC CANCER
GROWTH FAILURE IN PEDIATRICS
HEPATITIS B
HIV
INSOMNIA
JUVENILE IDIOPATHIC ARTHRITIS
MAJOR DEPRESSIVE DISORDER
MELANOMA
METASTATIC BREAST CANCER
MIGRAINE ATTACKS
MODERATE TO SEVERE PAIN
MULTIPLE SCLEROSIS
NON-HODGKIN'S LYMPHOMA
MULTIPLE MYELOMA
OVERACTIVE BLADDER
PAGET'S DISEASE
PARKINSON'S DISEASE
PLAQUE PSORIASIS
POSTMENOPAUSAL OSTEOPOROSIS
PREVENTION OF RISK OF PRETERM BIRTH
PRIMARY HYPERLIPIDEMIA AND MIXED DYSLIPIDEMIA
PROSTATE CANCER
PULMONARY ARTERIAL HYPERTENSION
RENAL CELL CARCINOMA
RHEUMATOID ARTHRITIS
SCHIZOPHRENIA
SMALL CELL LUNG CANCER
SPASMS AND COMPLEX PARTIAL SEIZURES
SYSTEMIC LUPUS ERYTHEMATOSUS
THROMBOCYTOPENIA
THYROID CANCER
TYPE 1 GAUCHER DISEASE
TYPE 2 DIABETES MELLITUS

***More than 160 New FDA Approved Indications for NMEs and BLAs from Jan. 2007 to May 1, 2011***

*Source: centerwatch.com*

**Efficient Upstream Process Development Programs Generated Using Early Molecule Assessment and Culture of Continuous Improvement**

Morris, A., Amgen

Originally presented on: 2/25/2013 10:30 - 11:00

To meet the tight timelines required for biopharmaceutical development Amgen has put considerable focus on improving the efficiency of upstream process development. These efficiencies begin with extensive analysis of protein expression and product quality at the molecule selection stage. This is followed with disciplined application of process platforms for clinical manufacturing. The results of this effort have resulted in processes that consistently produce commercial ready cell lines that meet aggressive clinical and commercial requirements.

**Past and Present Strategies, Challenges, and Benefits of Stable Cell Line Development at Genentech**

Shen , A., Genentech

Originally presented on: 2/25/2013 11:00 - 11:30

We have previously used a 2-cycle cell line development (CLD) approach (one line for early clinical production and a different line for late clinical and commercial production) to enable rapid entry into initial clinical studies. With the advance of new technologies, we have developed and implemented a 1-cycle CLD platform process that uses the same cell line from Tox all the way through commercial production, yet still allows fast entry into the clinic. This avoids potential comparability challenges associated with cell line changes and provides an opportunity to move from Phase I to Phase III quickly, if needed, for accelerating programs. This presentation will articulate cell line development strategies, throughputs, challenges, and benefits of past and current CLD processes.

**Streamlining the Discovery to Development Transition**

Allen, M., Pfizer

Originally presented on: 2/25/2013 11:30 - 12:00

State of the art expression technologies now allow routine production of recombinant protein therapeutics at levels adequate to support clinical testing and commercialization. However, identification of satisfactory production cell lines remains a tedious, resource costly, and time consuming process. Strategies to increase the efficiency of this process without compromising quality are clearly needed. To achieve more rapid cell line development we have taken a number of steps at Pfizer including 1) earlier engagement with Discovery organizations to harmonize production systems, 2) implementation of a molecule assessment to identify molecule liabilities at an early stage, and 3) implementation of site-specific integration technology to allow more consistent stable expression. These steps have enabled accelerated development; in one case resulting in the manufacturing of a master cell bank within three months of the identification of the final clinical candidate. This presentation will highlight the coordination of these technologies to allow accelerated development of protein therapeutics derived from diverse Discovery organizations and partners.

**Integration of Cell Culture Process Platform and Cell Line Expression Systems to accelerate Biopharmaceutical Development**

Kshirsagar, R., Biogen Idec

Originally presented on: 2/25/2013 12:00 - 12:30

Most biotech companies have a platform process for their various unit operations. We developed a cell culture process platform for inoculum expansion and fed-batch production bioreactor. Similarly, our colleagues in research developed a cell line development platform. However, the true potential of the two platforms was reached when they were integrated. The cell culture medium was used for host cell line adaptation. The clones were screened in the final fed-batch production process. This has allowed us to routinely deliver 5 g/L titer processes for early clinical programs. We have been able to further accelerate the development timelines by streamlining the development activities from transfection to clinical manufacturing by adopting high-throughput screening platforms and disposable technology.

**Enhancing Efficiency and Control for a Biopharmaceutical Manufacturing Process**

Mahtre, R., Biogen Idec

Originally presented on: 2/25/2013 13:35 - 14:05

Significant advances in the cell line and process technology have helped achieve high titer processes for therapeutic proteins. Downstream processing has also kept pace in enabling high recovery and purity of the proteins. These high yield processes however pose a challenge for maintaining process control. This presentation will cover technology developed to better understand and control processes so as to obtain consistent product quality. Techniques developed for screening raw materials, on line probes and other PAT tools will be presented.

Challenges and Innovative Approaches to Improving the Efficiencies of Process Development & Tech Transfer

**Gene Schaefer**
**API Large Molecule Process Development**

## Overview

- ABD #1 – looking back: efficient multi-site TT and S88-based process & information platform
- Integration of upstream and downstream development,
- Extension of a platform approach in manufacturing into development,
- Development of an overarching data and process recipe warehousing strategy to support development, tech transfer, criticality, and registrational filings
- Integrating an approach to criticality assessment with streamlining and automating process development activities,
- Summary – looking back, looking forward

## Overall Driver: Changing Face of Biotechnology!

Philip Sharp (left) & David Baltimore (center). Asilomar , 1975

*The use of recombinant DNA methodology promises to revolutionize the practice of molecular biology. While there has as yet been no practical application of the new techniques, there is every reason to believe that they will have significant practical utility in the future.*

P. Berg, Asilomar Summary Statement, 1975

- *Building biomanufacturing capacity—the chapter and verse.* M. Kamarck, Nature Biotech, 2006:
  - 1980s: "Just make it!"
  - 1990s: "Capacity Bottleneck"
  - 2000s: "It's a business"

## Structural/Performance Drivers for Organizational Change:

- ✓ Balance of resources across functional areas rarely exact match to project demands
- ✓ Conflicting priorities across projects, across functions
- ✓ Too many levels of "management"
- ✓ Protection of turf; failure to see larger picture
- ✓ Same old thinking; incremental changes
- ✓ Typically lack of definition around core vs non-core activities - "Business Model" definition lacking

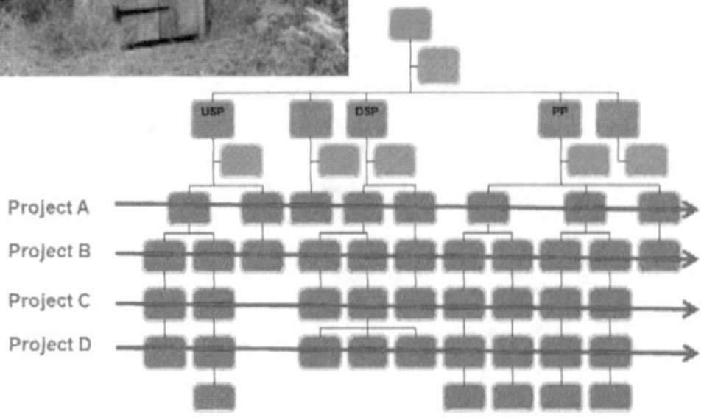

http://www.lifeinamatrix.com/

## Turn "Matrix" on its side

- ✓ **Reconfiguration** (transformational)
  - ✓ Major change; true innovation
  - ✓ Holistic approach, looking at bigger picture
  - ✓ Better alignment with core value: commercializing NMEs

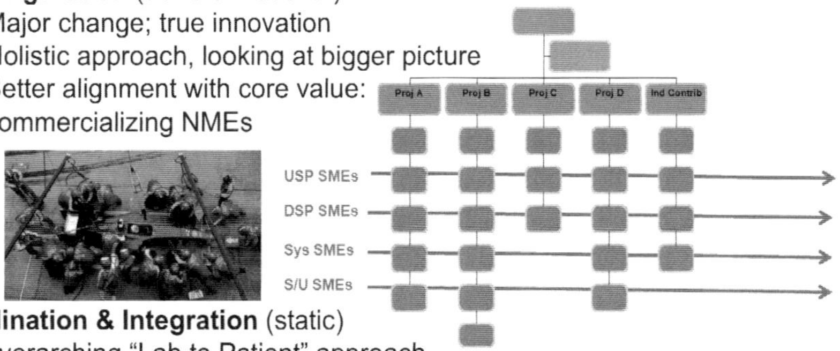


- ✓ **Coordination & Integration** (static)
  - ✓ Overarching "Lab to Patient" approach leveraging ANSI/ISA-95/88 (S95 & S88) standards: recipe-based context
  - ✓ Data management strategy
- ✓ **Learning Strategies** (dynamic)
  - ✓ Significant Cross-Training; create development modules & toolboxes
  - ✓ Cross-project Heavyweight/Lightweight teams
  - ✓ Oversight by USP/DSP SMEs

janssen

## Leveraging Recipe Framework (S88/S95)

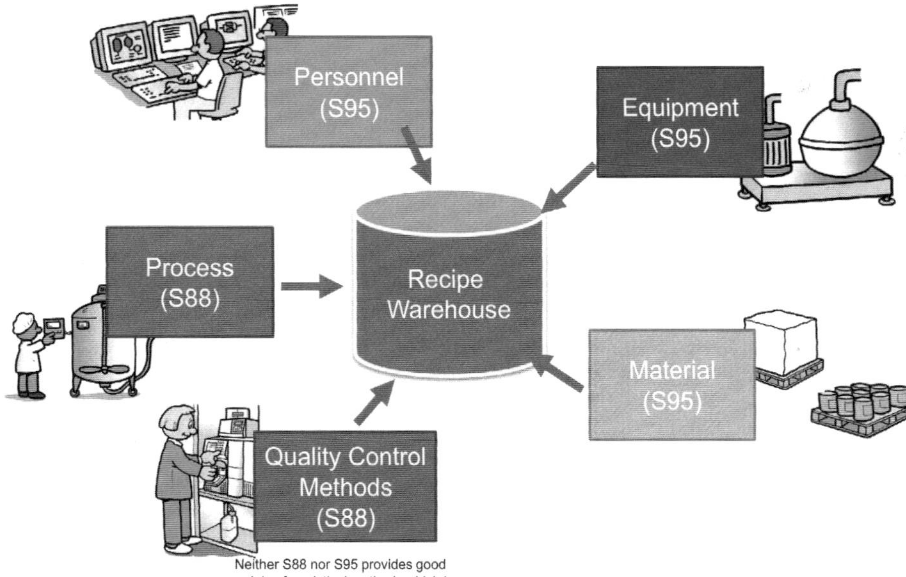


Neither S88 nor S95 provides good models of analytical methods which is important for drug development (J&J)

janssen

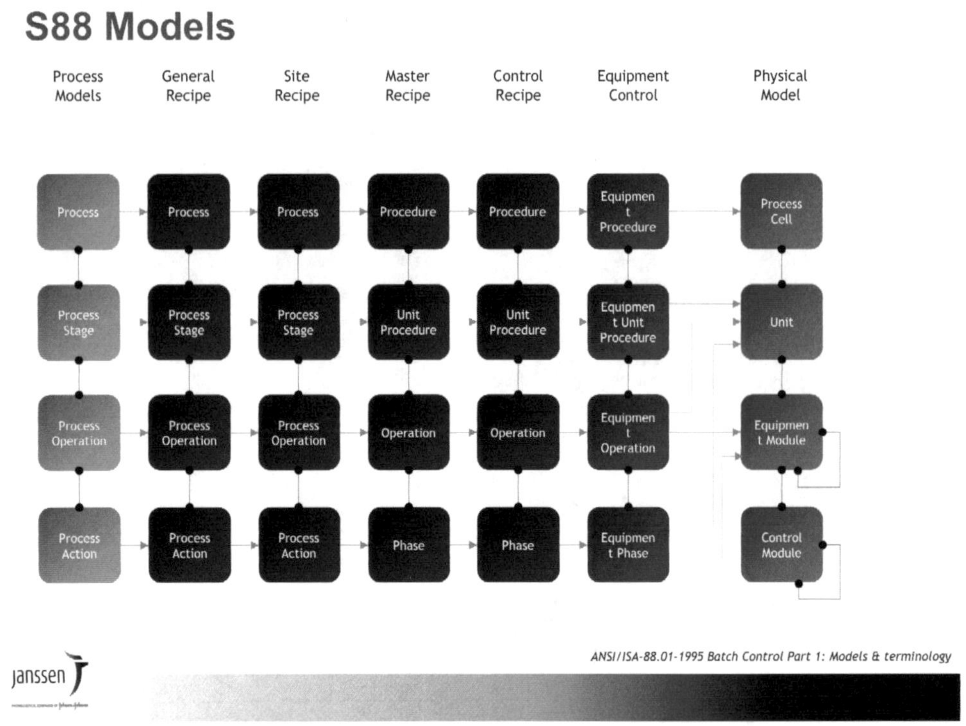
S88 Models
Process Models
General Recipe
Site Recipe
Master Recipe
Control Recipe
Equipment Control
Physical Model
Process
Process
Process
Procedure
Procedure
Equipment Procedure
Process Cell
Process Stage
Process Stage
Process Stage
Unit Procedure
Unit Procedure
Equipment Unit Procedure
Unit
Process Operation
Process Operation
Process Operation
Operation
Operation
Equipment Operation
Equipment Module
Process Action
Process Action
Process Action
Phase
Phase
Equipment Phase
Control Module
ANSI/ISA-88.01-1995 Batch Control Part 1: Models & terminology
janssen

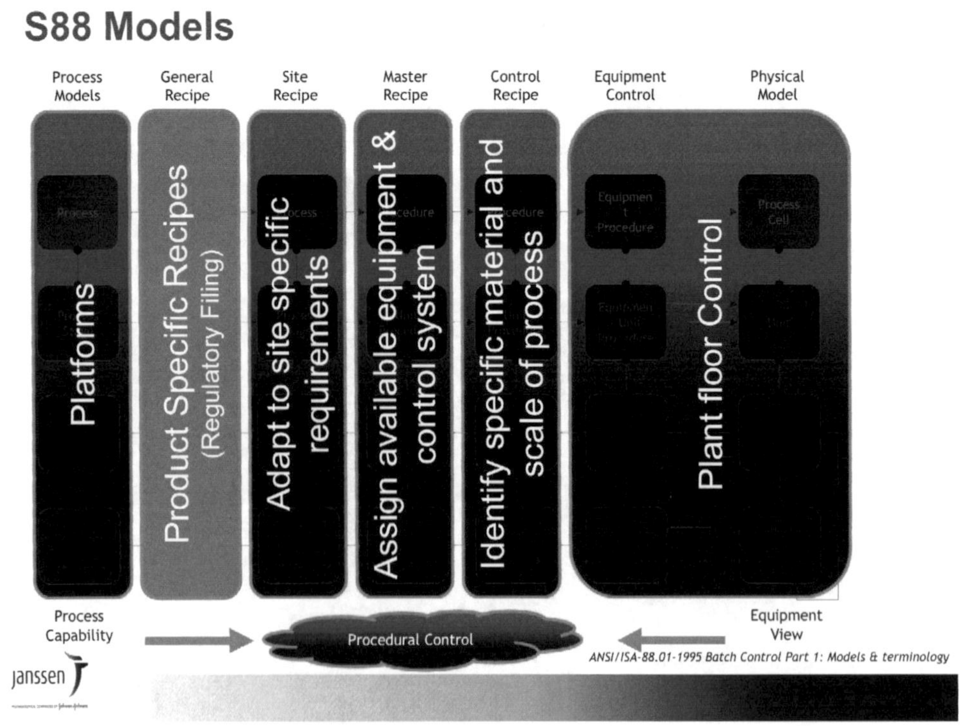
S88 Models
Process Models
General Recipe
Site Recipe
Master Recipe
Control Recipe
Equipment Control
Physical Model
Platforms
Product Specific Recipes
(Regulatory Filing)
Adapt to site specific requirements
Assign available equipment & control system
Identify specific material and scale of process
Plant floor Control
Process Capability
Procedural Control
Equipment View
ANSI/ISA-88.01-1995 Batch Control Part 1: Models & terminology
janssen

## Recipe Framework Implications

- Recipes modeled independent of scale
- Platformed to enable scientific learning's to be transcribed via Criticality Analyses (CA)

- **P** Production Bioreactor
  - **S** Initialize
  - **S** Equilibrate
  - **S** Inoculate Bioreactor
  - **S** Continuous Perfusion
  - **S** Inactivate Bioreactor

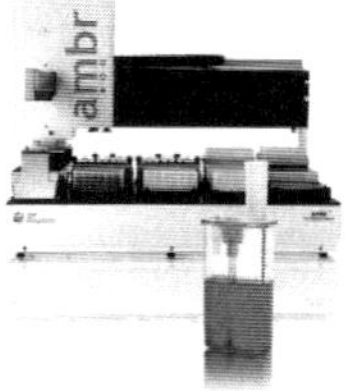


Lab Automation
~15mL

Lab Scale
~3-50 L

Plant Scale
~500-1000 L

janssen

## Recipe Framework Implications

- Recipes modeled independent of scale
- Platformed to enable scientific learning's to be transcribed via CA

- **P** Production Bioreactor
  - **S** Initialize
  - **S** Equilibrate
  - **S** Inoculate Bioreactor
  - **S** Continuous Perfusion
  - **S** Inactivate Bioreactor

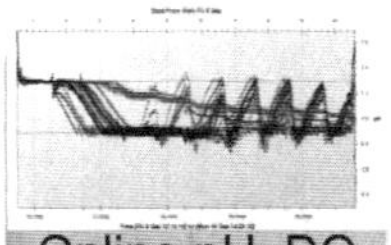

Online pH, DO

Cedex Novaflex

Off Line
Raman

Most Analytical Tools are Scale Independent

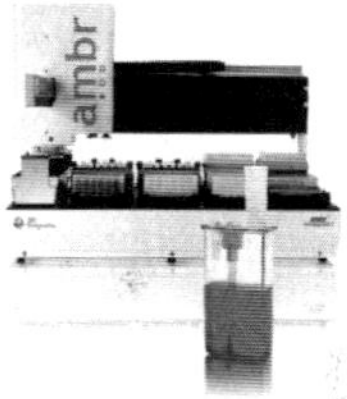


Lab Automation
~15mL

Lab Scale
~3-50 L

Plant Scale
~500-15,000 L

janssen

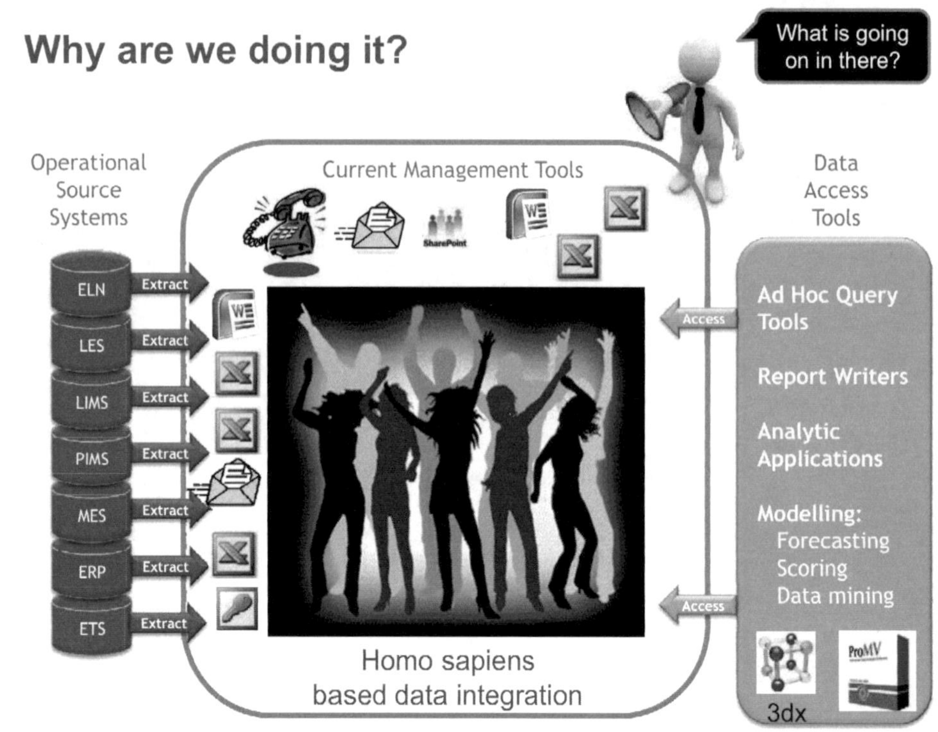


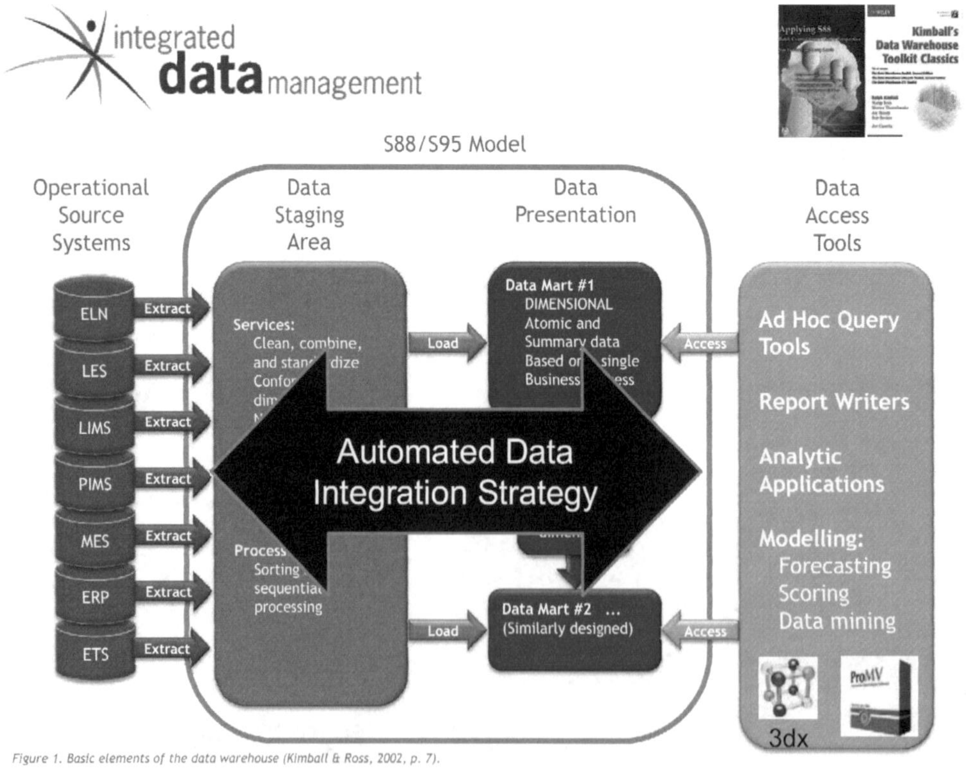


*Figure 1. Basic elements of the data warehouse (Kimball & Ross, 2002, p. 7).*

## From data warehousing to information management

- Organization = enabler of knowledge creation & application
- Organizational capabilities = clusters of knowledge sets & routines that are translated into distinctive activities
- Embedded knowledge = combination of tacit & explicit
- Need a knowledge management strategy
- Survival in a rapidly-changing knowledge environment drives a need to continuously adapt knowledge base and capabilities: combination of **internal learning** (learning by doing) and **external learning** (assimilated by imitation, licensing, acquisition, or inter-organizational collaboration)
- Two learning processes are not substitutes but interdependent

Santos, F., Technol Forecasting & Social Change **70**: 687-715 (2003)

janssen

## Organizational Knowledge Building: Worst (typical) case

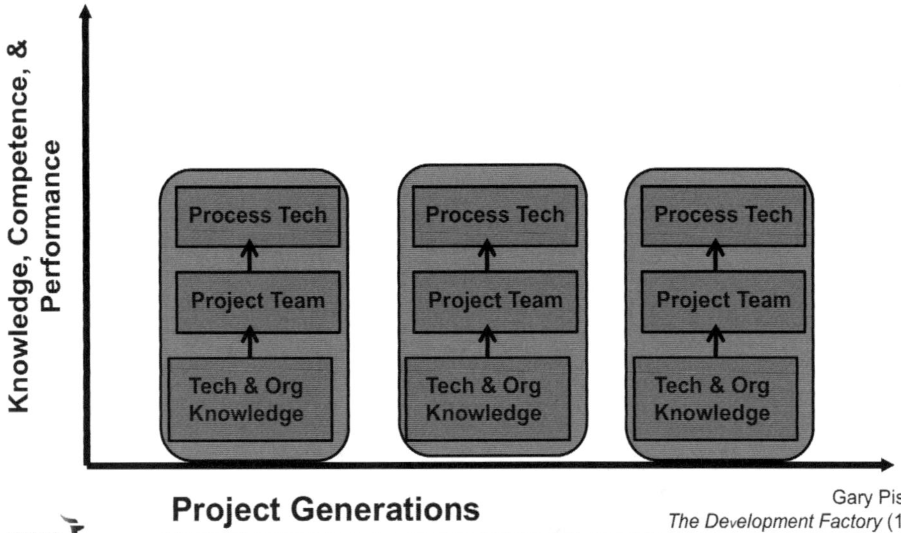


Gary Pisano, *The Development Factory* (1997)

janssen

## Organizational Knowledge Building: Desired State

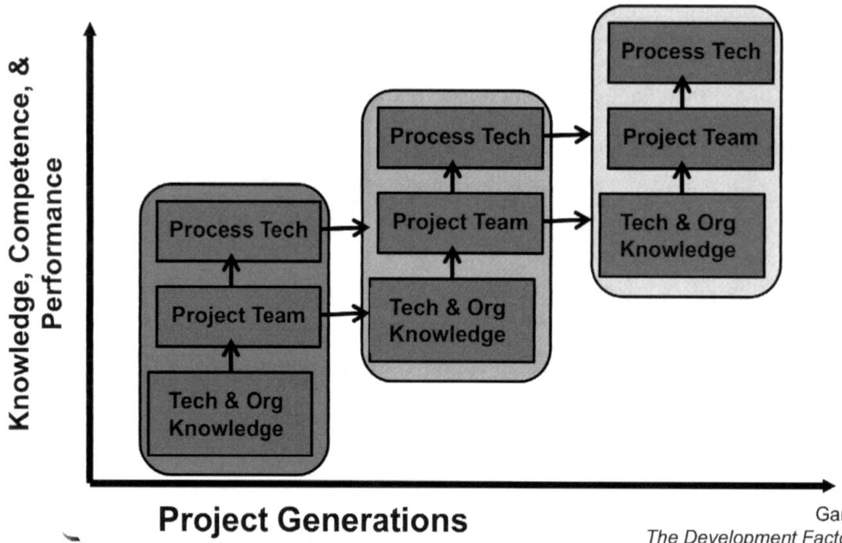


Gary Pisano, *The Development Factory* (1997)

Janssen

## Create "Platforms" for Development Process

### Common Development Modules

- Ensures teams follow similar paths in terms of general experimental approach
- Presents opportunities to leverage lessons learned across projects

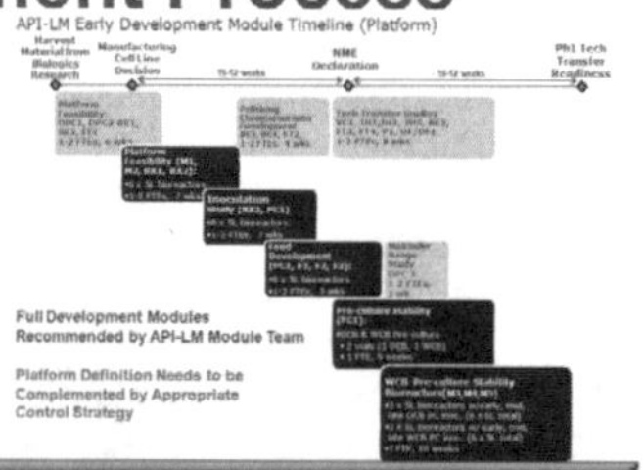


### Building a Knowledge Based Organization

#### Common Recipe Approach

- Ensures that we all "speak the same language" promoting consistency
- Presents a significant opportunity to realize efficiencies across our systems landscape as medicines moves through their approval lifecycle

#### Common Data Warehouse

- Enables leveraging of information across systems to help drive data driven decisions
- Presents opportunity to link across systems and leverage said knowledge during clinical manufacturing, tech transfer, and filings

## Reaching our Future - Integrated Data Management

**People – Leveraging skill sets**

Leading Change Teams – Heavyweight & Lightweight Teams

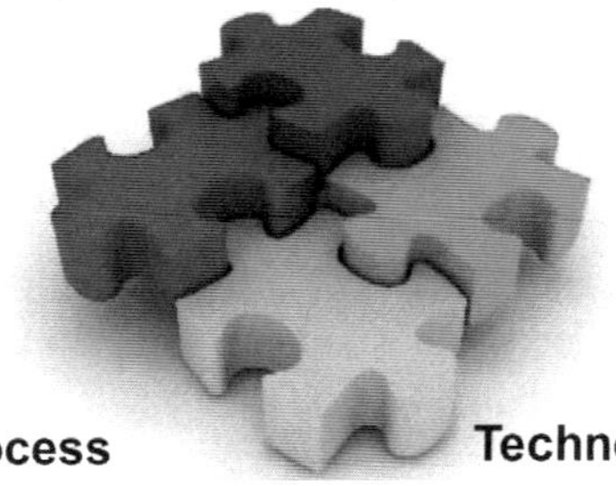

**Process**

- Integrate experimental execution platforms through modules, toolboxes, & common data structures (S88/S95)
- Governance mechanisms established to balance R&D versus commercial influence

**Technology Implementation**

- Leverage Standard technology platforms where possible, *e.g.* OSI-PI, Werum, DeltaV, Symyx eLN, Freeslate
- Provide standard data transfer protocols between platforms (S88/S95)

janssen

## MVA & Latent Variables?? What's that?

- From a physical, chemical, or biological perspective, pharmaceutical products and processes are complex multi-factorial systems.

Inputs Variables (X)

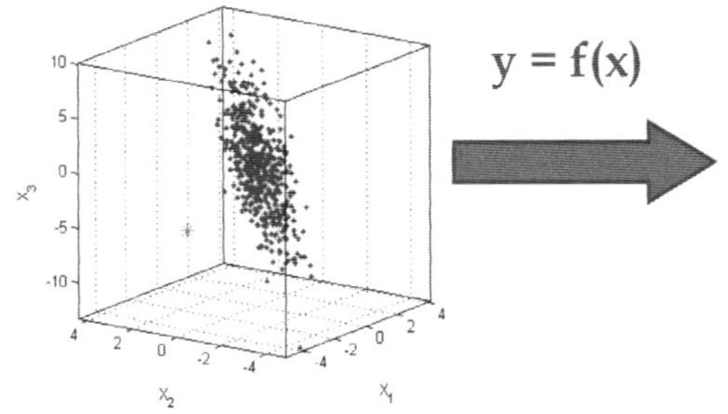


$y = f(x)$

Inputs to the process control the variability of the output

**Latent variable space**

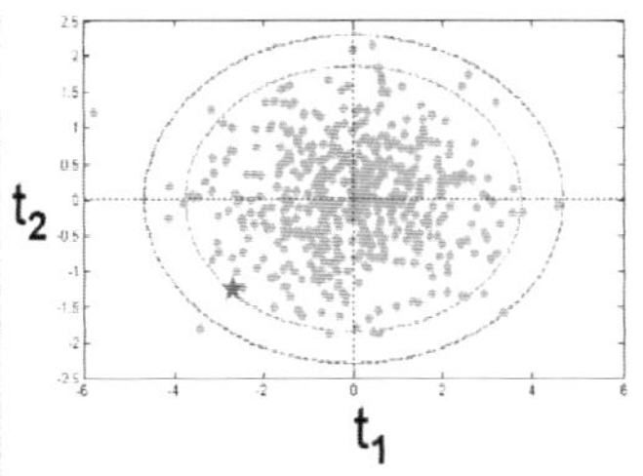


Help reduce the dimensionality of data to help in analysis

janssen

## Analytics Dashboard

3dx

- A versatile tool proving scientists and engineers plug and play options for data visualization was leveraged from informatics organization
  - Internally developed tool called Three Dimensional Explorer (3dx)
  - Modular analytics dashboard that can quickly integrate to external data sources and computational tools
- Leveraged Prosenus Multivariate Data analysis libraries to provide MVA capabilities
  - Casual users can view data
  - Advanced users can build models that can be viewed by others

## Calculation library design

- System architecture diagram:

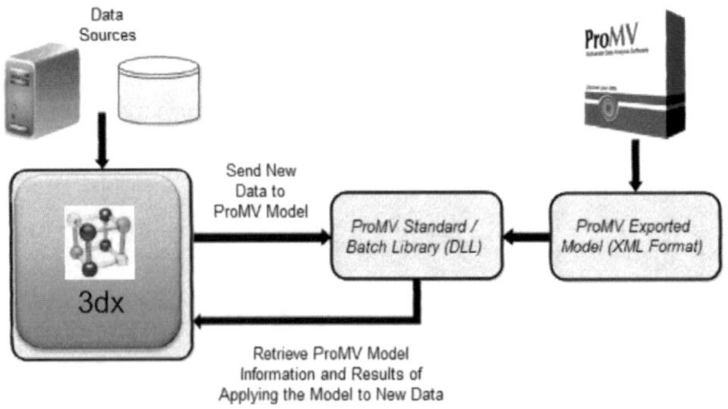


  - PCA/PLS models built in state-of-the art desktop software, ProSensus MultiVariate (ProMV)
  - Models exported in open XML format
  - Libraries embedded in 3dx via DLL's

## Calculation library design

- System architecture diagram:

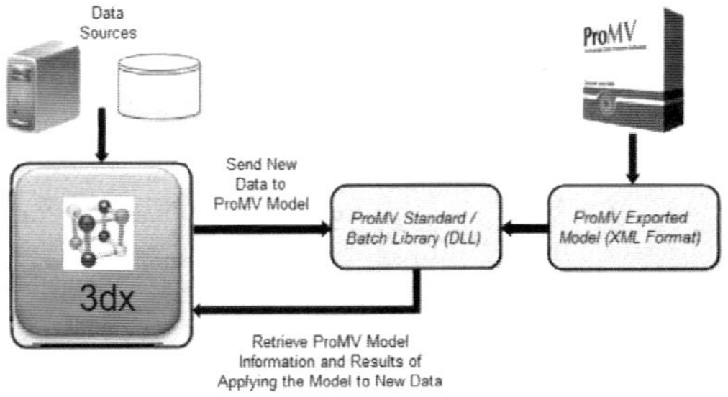


  - PCA/PLS models built in state-of-the art desktop software, ProSensus MultiVariate (ProMV)
  - Models exported in open XML format; libraries embedded in 3dx DLL's
  - 3dx calls methods in ProMV Libraries as needed, passing new data and specifying model to use
  - ProMV Libraries operate on data and model, return results to 3dx

## Interactive Analytics Dashboard

3dx

- End users now have a fully integrated analytics dashboard
  - Standard viewers to profile data
  - Enhanced integration for MVA portal
  - Ability to pull multiple data sets for criticality analyses
- Capabilities to build out standardized reports
- Ability to transfer models to control systems

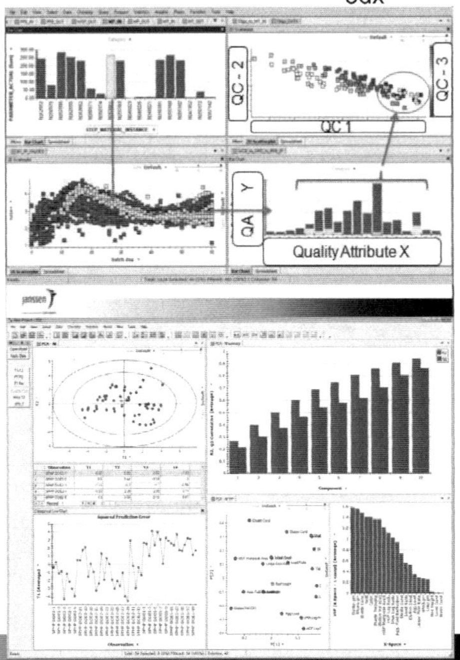

## Viable Cell Density versus Media Lot

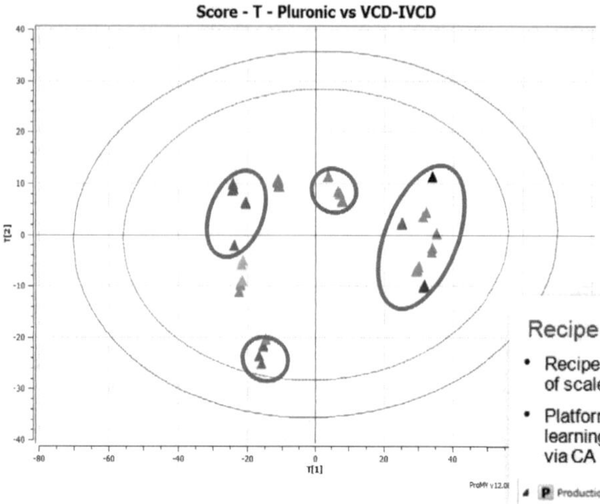


Clusters of results in SCORE plot driven by raw material lot # in platform medium – across multiple scales, sites, and campaigns

### Recipe Framework Implications

- Recipes modeled independent of scale
- Platformed to enable scientific learning's to be transcribed via CA

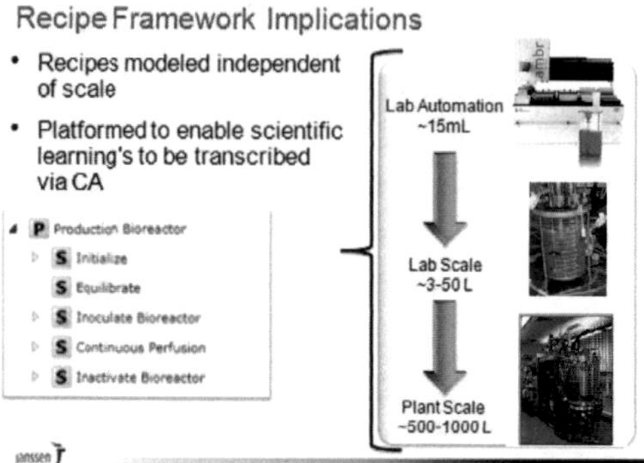


## Purification DoE Studies

3dx

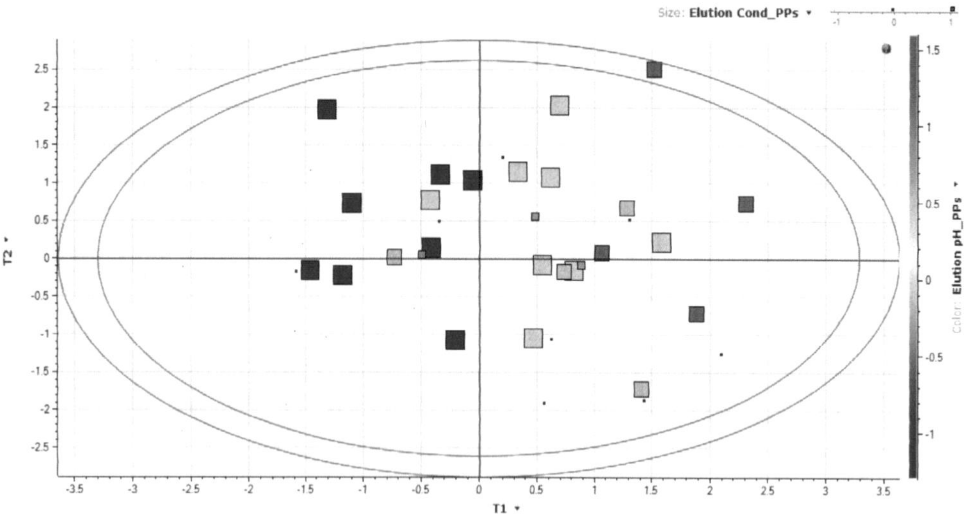

## Criticality/Control strategy derived from either univariate or multivariate analyses

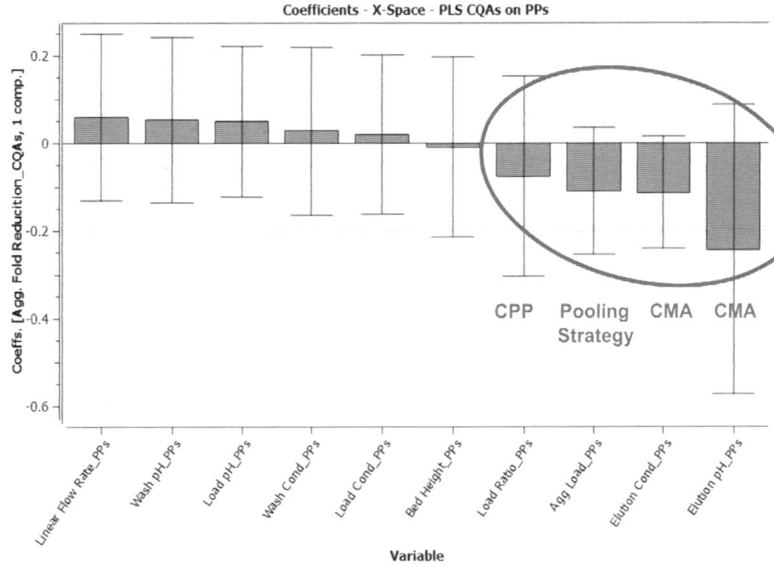


Janssen

## Riding the Wave of Innovation in Biotech

## Summary

3dx

ProMV

- ✓ ABD #1 – **making progress**
- ✓ Integration of upstream and downstream development – **working well**
- ✓ Extension of a platform approach in manufacturing into development – **significant progress**
- ✓ Development of an overarching data and process recipe warehousing strategy to support development and registrational filings – **significant progress**
- ✓ Integrating an approach to criticality assessment with streamlining and automating process development activities – **moving ahead**

janssen

## Riding the Wave of Innovation in Biotech

## Acknowledgements

- Paul McKenzie
- Adam Fermier
- Steve Mehrman
- John Cunningham
- Fran Meacle
- Tony Lubiniecki
- Andrew Skalkin, Victor Lobanov, & Walter Cedeno
- Many others who have helped drive the strategy and implementation within Janssen
- Our ProSensus collaborators especially John MacGregor, Mark-John Bruwer, & Mark Mordeca

Statements of fact and opinions expressed are those of the participants individually and, unless expressly stated to the contrary, are not the opinion or position of Janssen Research & Development LLC or its affiliates.

janssen

**Using Cell Culture Platform and High-Throughput Bioreactors to Streamline Upstream Process Development**

Schmelzer, A., MedImmune

Originally presented on: 2/25/2013 14:35 - 15:05

With the ever increasing cost of developing biopharmaceuticals and number of drug candidates in the R&D portfolio, more efficient systems are required to progress molecules quickly through the R&D pipeline into clinical development and commercial launch. Over the past decade, MedImmune has developed expertise to express monoclonal antibodies at titers greater than 10 g/L at manufacturing scale and has leveraged these advances into the design of our platform processes. In addition to platform cell culture processes that yield high titers, the implementation of high-throughput bioreactor systems, including ambr (by TAP Biosystems), has been instrumental in efficiently defining a commercially-viable process early on with minimal resources. In this work, we demonstrate how we have leveraged these tools for (1) accelerating clone selection, (2) Phase 1 clinical trial and commercial process design, and (3) scale-down model definition suitable for design space, QbD, or other BLA-enabling activities.

**Speeding Whole Bioprocess Solutions: the Crucial Interface Linking the Bioreactor to Purification**

Hoare, M., University College London
Originally presented on: 2/25/2013 15:05 - 15:35

The primary recovery sequence is that which helps link the bioreactor output to the high resolution purification and formulation stages. The operations involved in this sequence are amongst the most challenging for successful predictive design to help take us from the bench to the manufacturing scale. Early decisions taken on the preferred options to be used in the primary recovery stages and predicting how these choices best function together will help us to:- (a) Determine ways in which to run the bioreactor to help deliver a robust whole bioprocess of an acceptable performance. (b) Select purification sequences which are robust to variations in feed from the upstream processing stages. The paper will discuss the integration of available and new ultra scale-down tools for the study at the early stage of process development of different primary recovery options. A range of next generation antibody-related products from both mammalian and bacterial cell hosts will be addressed. These pose challenging bioprocessing problems ranging from high sensitivity to process shear stress through to the need to incorporate selective broth conditioning stages.

**Upstream and Downstream Strategies to Actively Control Product Quality Attributes**

Pettit, D., Amgen

Originally presented on: 2/25/2013 15:35 - 16:05

Historically, the strategy for large molecule process development has been centered on the idea that reducing process and raw material variability will lead to products with minimal variability in their quality attributes. In this scenario, the purpose of in-process and final product release testing is to assess and control process variability. In recent years, greater understanding of specific attribute criticality has allowed us to envision processes specifically designed to control critical product quality attributes. This presentation will describe examples where product quality attributes were well understood and strategies for upstream and downstream process design resulted in the production of molecules with appropriate, pre-defined product quality attributes.

**Analytical Strategies Enabling Expedited Development of Therapuetic mAbs**
Dufield , R., Pfizer
Originally presented on: 2/25/2013 16:45 - 17:15

A common challenge faced in biopharma is the ability to advance numerous projects throughout development while minimizing cost and timelines and maximizing the robustness of the CMC strategy. The portfolios at many biotech companies include multiple therapeutic monoclonal antibodies (mAbs) along with several other therapeutic modalities in development. With the number of monoclonal antibody programs increasing and rapidly advancing through development, it is imperative to apply state-of-the-art analytics and to implement streamlined/efficient work processes while maintaining product quality and safety. At Pfizer we have implemented a platform analytical approach to rapidly advance our therapeutic mAb programs. This analytical approach monopolizes on the structural homology across mAbs yet provides modular options to address mAbs with unique characteristics. State-of-the-art analytics (e.g. Ultrahigh-resolution ESI-QTOF mass spectrometry to confirm sequence fidelity during clone selection) and streamlined work processes have enabled the rapid advancement of our monoclonal antibody therapeutics. This presentation will describe our mAb platform analytical approach and organizational work process towards methods development, methods qualification, and product characterization to facilitate the rapid advancement of mAb therapeutics in the portfolio.

# VLP-based human vaccines: Production and analysis

**Qinjian Zhao**

*School of Public Health, Xiamen University, China*

*Feb 25, 2013 @ AccBio/QbD*

# Outline

- VLP based vaccines: HBV, HPV and HEV
- **Platforms** for recombinant protein production
- First licensed HEV vaccine – Hecolin®
- **VLP analysis**: process development, product licensure and post-licensure comparability

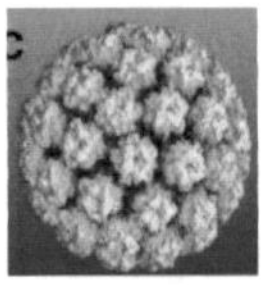

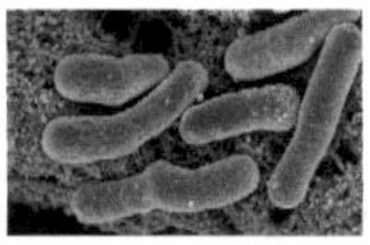

# Outline

- VLP based vaccines: HBV, HPV and **HEV**
- **Platforms** for recombinant protein production
- First licensed HEV vaccine – Hecolin®
- **VLP analysis**: process development, product licensure and post-licensure comparability

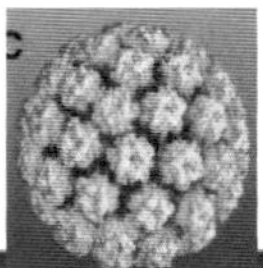

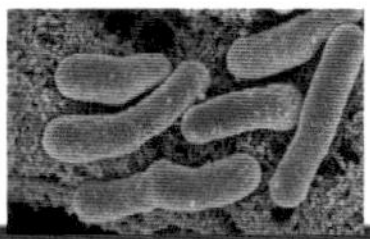

National Institute of Diagnosis and Vaccine Development in infectious diseases, Xiamen U

## Expression hosts for therapeutic proteins

- 130 licensed TP products in US/EU
- Among 48% were made in microbial cells, 34% were made in *E coli* and 13% in yeast.

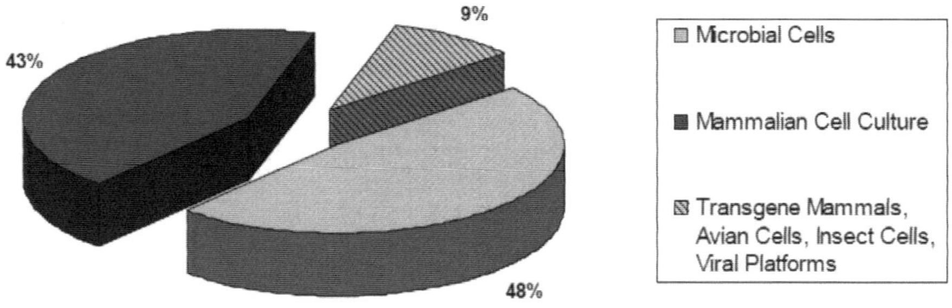

# Recombinant VLP based vaccines

- HBV - RECOMBIVAX HB® 22 nm S. cerevisiae
- HBV - Engerix ® 22 nm S. cerevisiae
- HPV - GARDASIL® 40-60 nm S. cerevisiae
- HPV – Cervarix® 40-60 nm Baculovirus

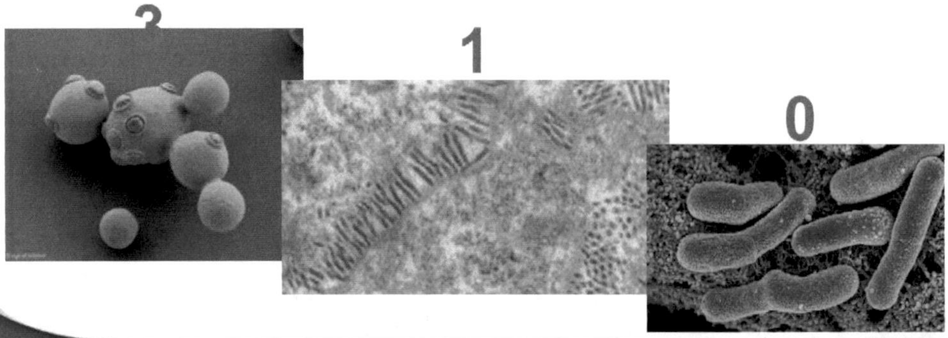


National Institute of Diagnosis and Vaccine Development in infectious diseases, Xiamen U

# Complex VLP structure - HBsAg

- 22 nm lipid (~30%)-protein VLP
- Octahedral structure
- 96 monomers per VLP
- MW ~3,000,000 Da

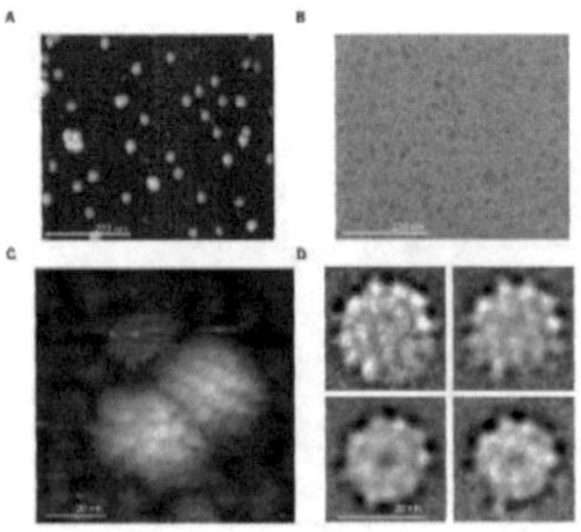

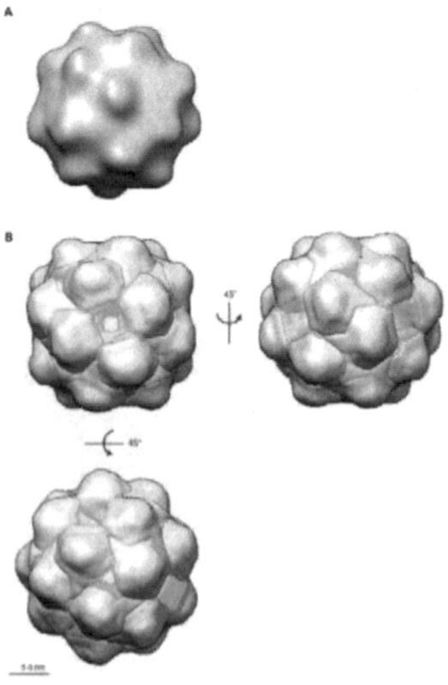

Mudler *et al.* (2012). PLoS ONE, 10.1371/journal.pone.0033235

# Complex VLP structure - HPV L1

- 40-60 nm VLP
- Icosahedral (T=7) structure
- 72 pentamers or
  360 monomers per VLP
- MW ~20,000,000 Da

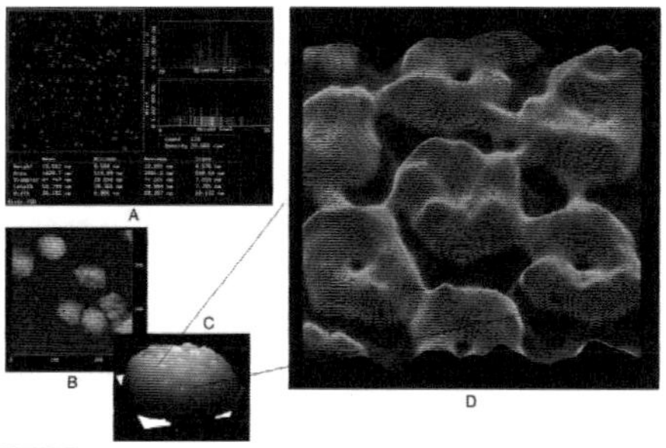


Zhao *et al.* (2012), Nanomedicine, 8, 1182-1189.

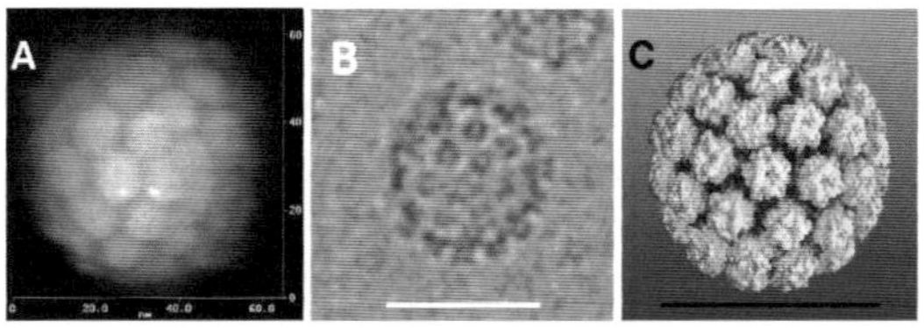


Zhao *et al.* (2012). Virol. J., 9:52 (doi:10.1186/1743-422X-9-52).

National Institute of Diagnosis and Vaccine Development in infectious diseases, Xiamen U

# Production platforms for recombinant VLP based vaccine

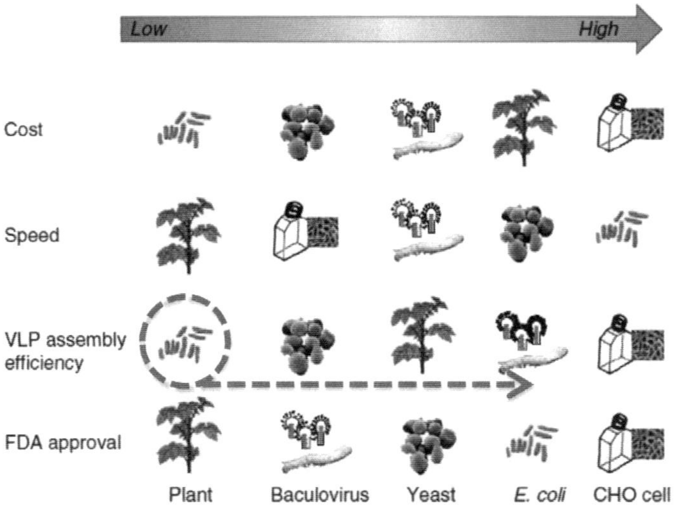


Cho, H. J., Y. K. Oh, et al. (2011). "Advances in human papilloma virus vaccines: a patent review." Expert Opin Ther Pat.

National Institute of Diagnosis and Vaccine Development in infectious diseases, Xiamen University

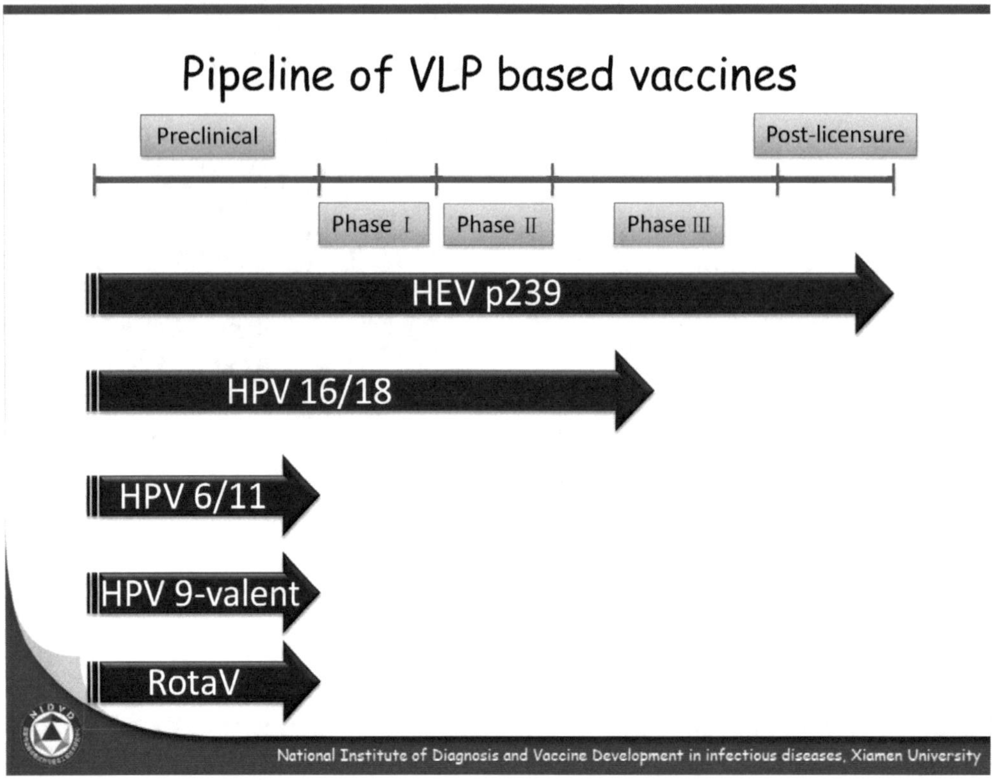
Pipeline of VLP based vaccines
Preclinical
Post-licensure
Phase I
Phase II
Phase III
HEV p239
HPV 16/18
HPV 6/11
HPV 9-valent
RotaV
National Institute of Diagnosis and Vaccine Development in infectious diseases, Xiamen University

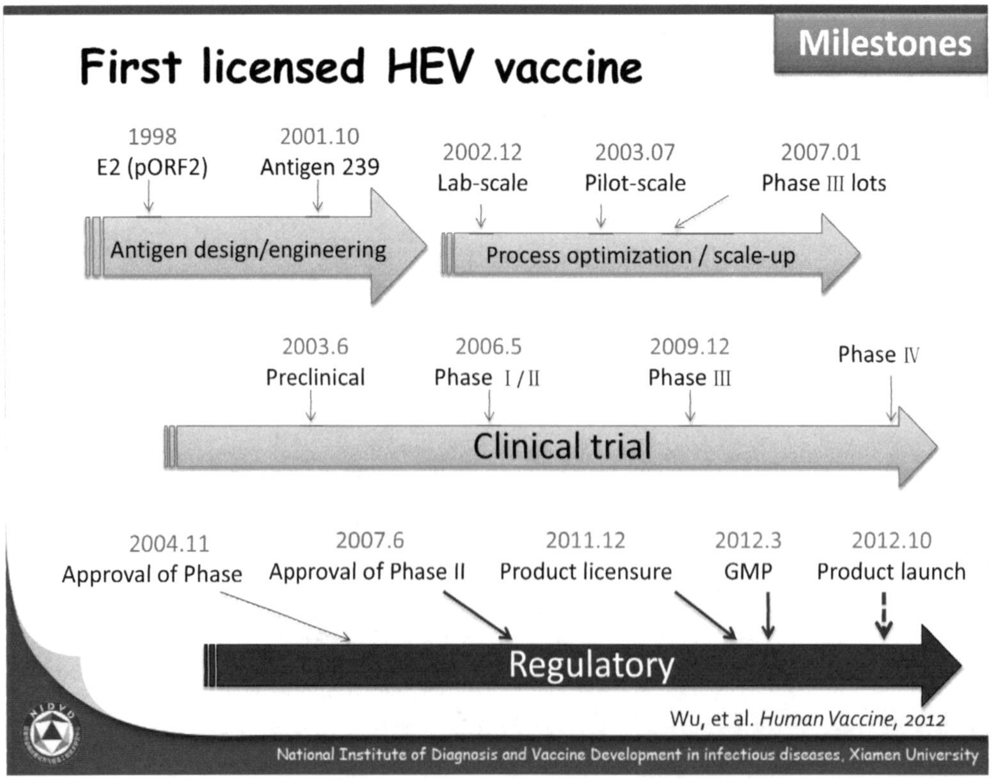
First licensed HEV vaccine
Milestones
1998
E2 (pORF2)
2001.10
Antigen 239
Antigen design/engineering
2002.12
Lab-scale
2003.07
Pilot-scale
2007.01
Phase III lots
Process optimization / scale-up
2003.6
Preclinical
2006.5
Phase I/II
2009.12
Phase III
Phase IV
Clinical trial
2004.11
Approval of Phase
2007.6
Approval of Phase II
2011.12
Product licensure
2012.3
GMP
2012.10
Product launch
Regulatory
Wu, et al. Human Vaccine, 2012
National Institute of Diagnosis and Vaccine Development in infectious diseases, Xiamen University

# 14-year development cycle for HEV vaccine

COMMENTARY

Human Vaccines & Immunotherapeutics 8:6, 1-5; June 2012; © 2012 Landes Bioscience

## Hepatitis E vaccine development

A 14-year odyssey

Ting Wu,[1,2,3] Shao-Wei Li,[1,2,3] Jun Zhang,[1,2,3] Mun-Hon Ng,[1,2] Ning-Shao Xia[1,2,3] and Qinjian Zhao[1,2,*]

[1]National Institute of Diagnostics and Vaccine Development in Infectious Diseases; Xiamen University; Xiamen, China; [2]School of Public Health; Xiamen University; Xiamen, China; [3]School of Life Sciences; Xiamen University; Xiamen, China

The first prophylactic vaccine, Hecolin®, against hepatitis E virus (HEV) infection and the HEV associated with the most recent one in Uganda in 2007–2010[5] and with the largest outbreak taking place in Xinjiang Province, China

National Institute of Diagnosis and Vaccine Development in infectious diseases, Xiamen U

# R & D of HEV239 vaccine

1998 E2 (pORF2) → 2001.10 Antigen 239 → Antigen design/engineering

2002.12 Lab-scale → 2003.07 Pilot-scale → 2007.01 Phase III lots → Process optimization / scale-up

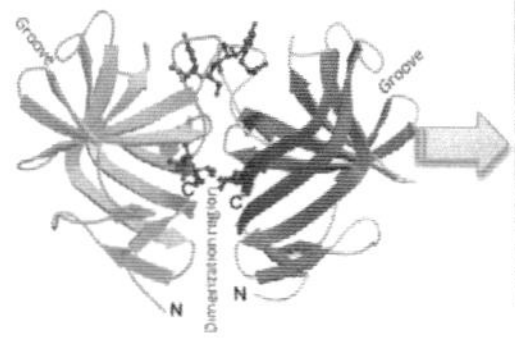

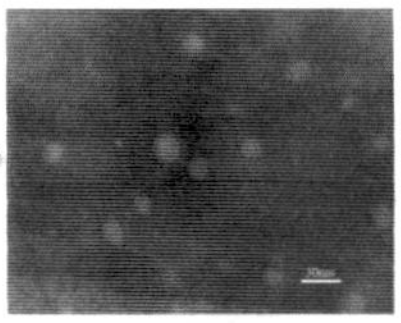

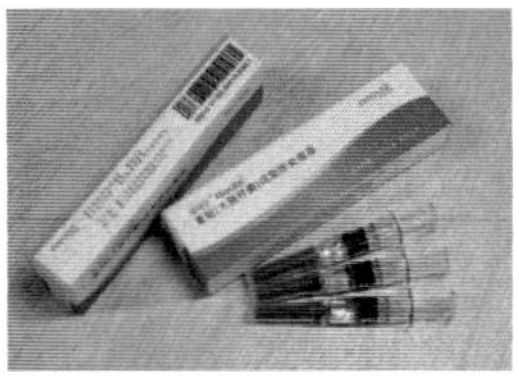

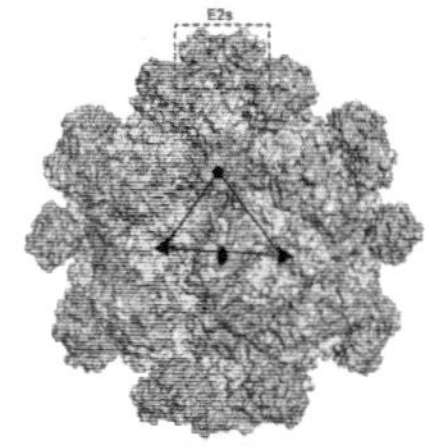

T=1 HEV VLP (PDB No.2ZTN)

National Institute of Diagnosis and Vaccine Development in infectious diseases, Xiamen University

## Much preferred immuno-reactivity of patient sera to E2 (over monomeric form)

- pORF2 forms dimer, or higher order assemblies upon over expression
- Sucessful expression of HEV pORF2 fragment, E2, in *E. coli*
- Lost immuno-reactivity of E2 dimer upon denaturation into monomer

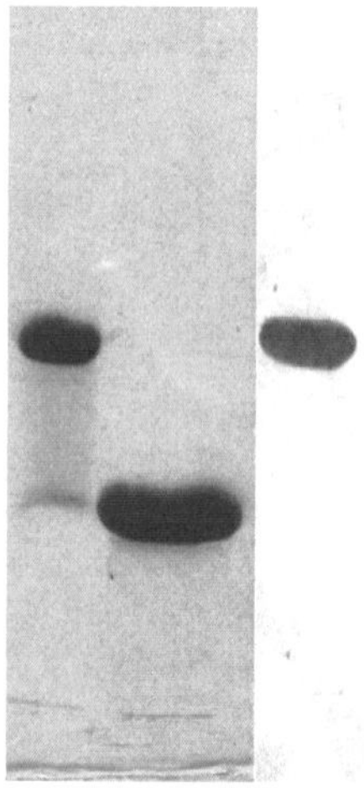

Western Blot

Zhang et al. *J Med Virol* 2001.

National Institute of Diagnosis and Vaccine Development in infectious diseases, Xiamen U

## Dimerization of HEV pORF2

(E2s domain-149aa)

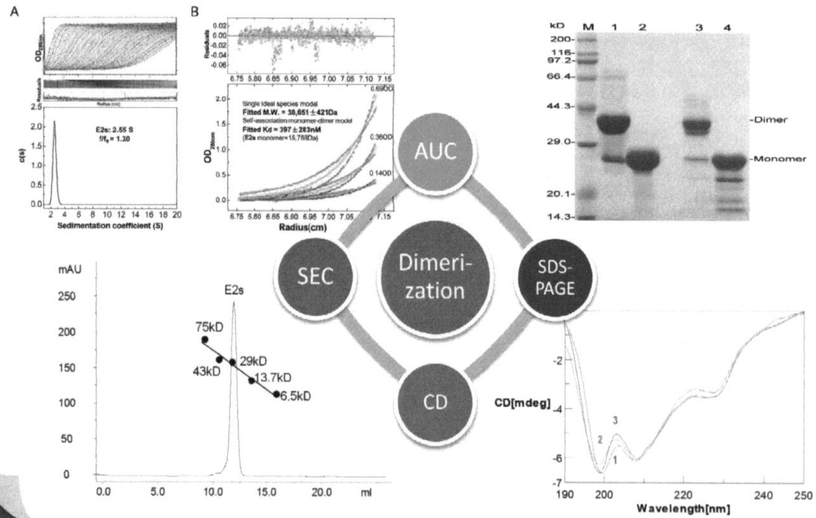


Li, et al. *PLoS Pathogen*, 2009.

National Institute of Diagnosis and Vaccine Development in infectious diseases, Xiamen University

# Structure of HEV pORF2
## (E2s domain-149aa)

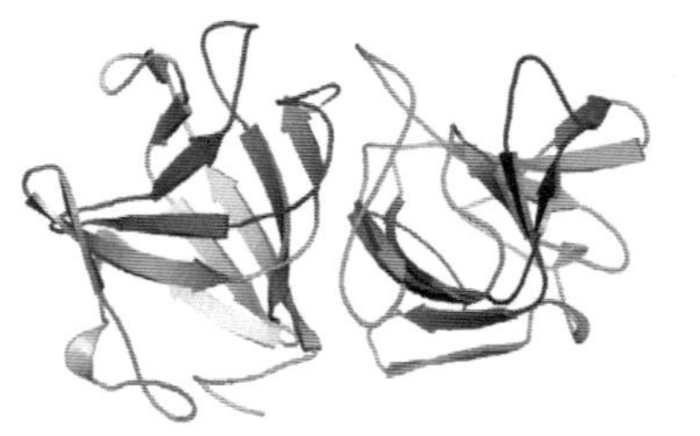

- Typical capsid protein folding: **β-barrel**
- Dimerization: strong **hydrophobic interactions**
- Groove region: **flexible loops**

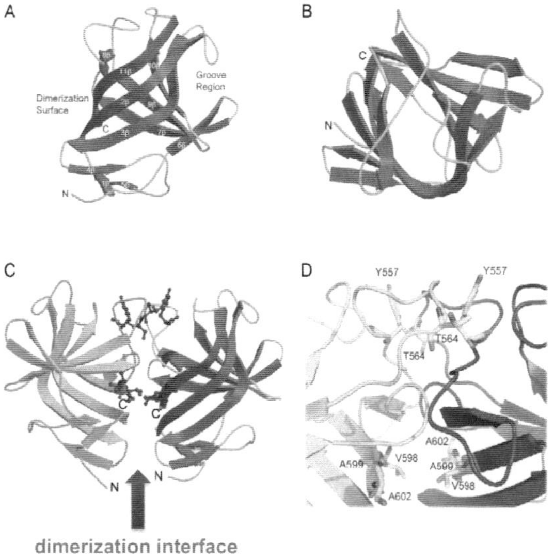


Li et al. *PLoS Pathogens*. 2009



# Mapping of key residues on E2s

✓ Dimer interface

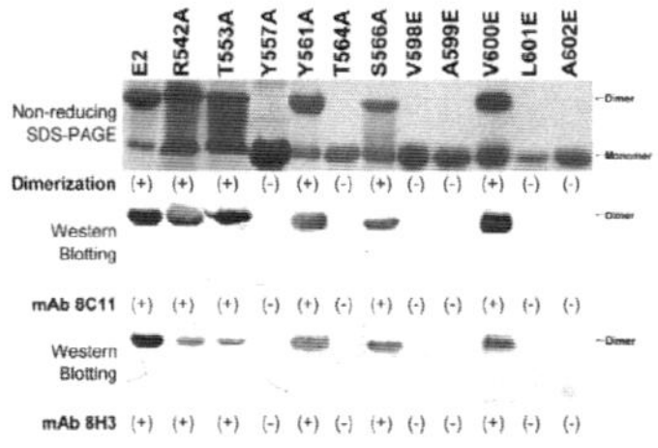


✓Immuno dominant neutralizing sites

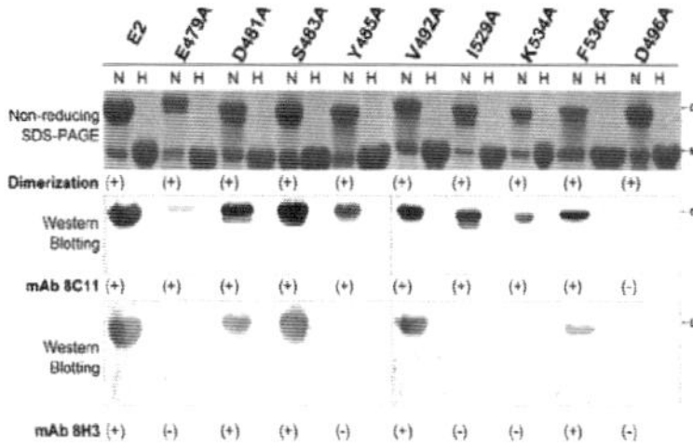


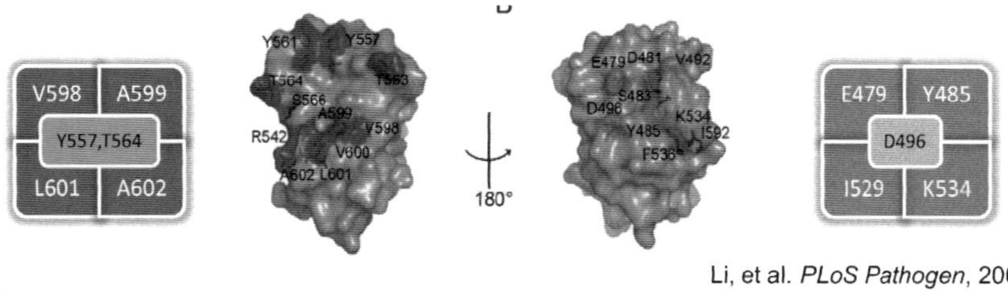


Li, et al. *PLoS Pathogen*, 2009.

# Structure of E2s:8C11 mAb complex

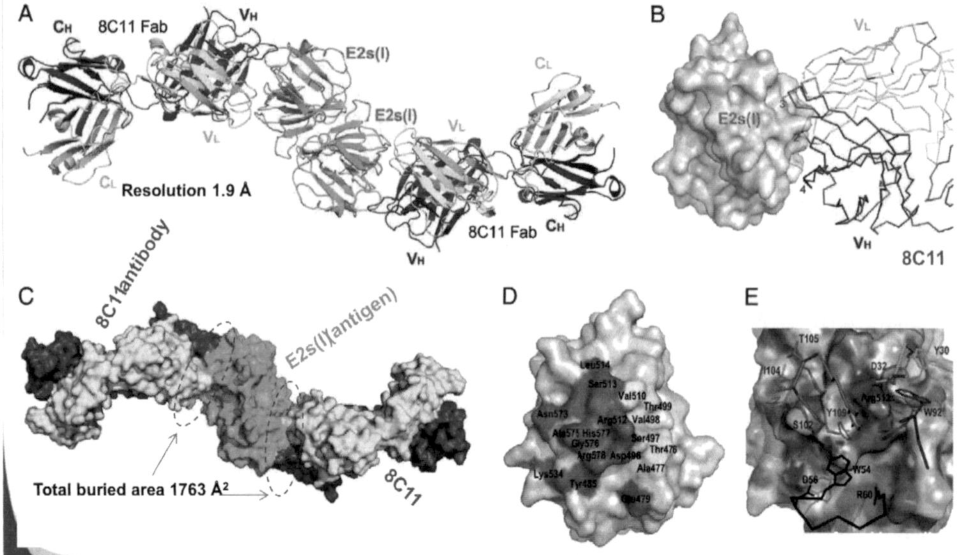


8C11 epitope on E2s: Asp496-Thr499, Val510-Leu514 and Asn573-Arg578

Tang *et al., PNAS.* 2011.



# Enhancing immunogenicity by epitope multiplicity

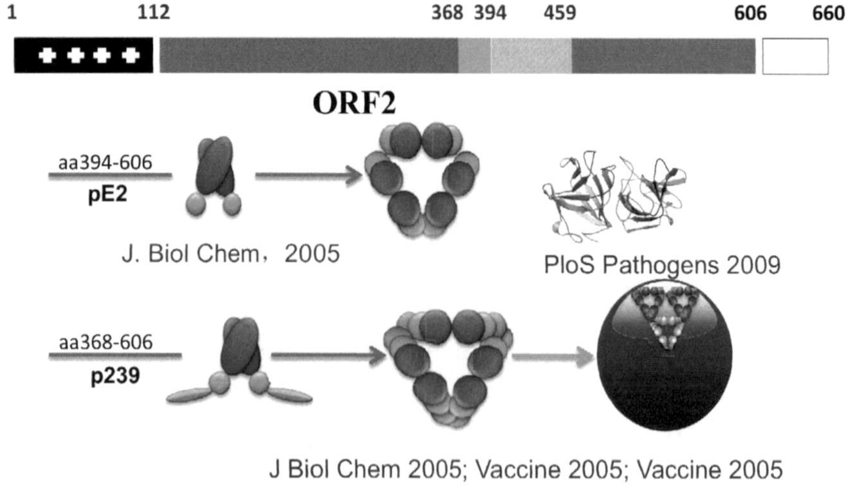


J. Biol Chem，2005

PloS Pathogens 2009

J Biol Chem 2005; Vaccine 2005; Vaccine 2005

National Institute of Diagnosis and Vaccine Development in infectious diseases, Xiamen University

# Enhancing immunogenicity by making multimeric VLP antigen

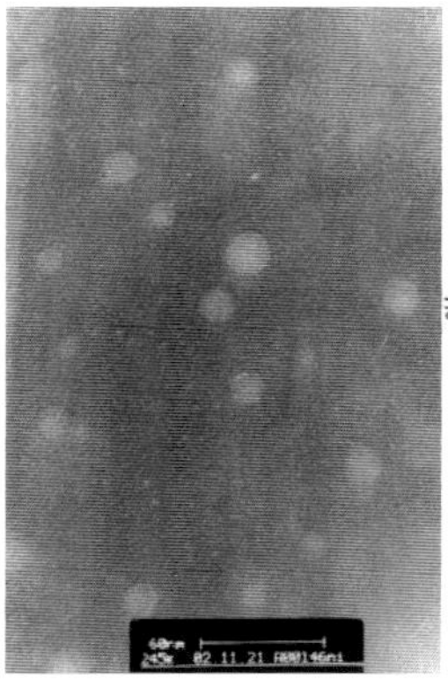

Electron microscope

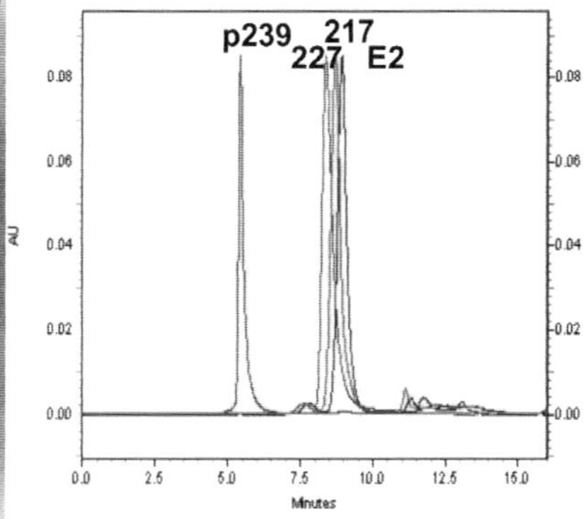


Gel filtration by HPLC (GSW3000xl column)

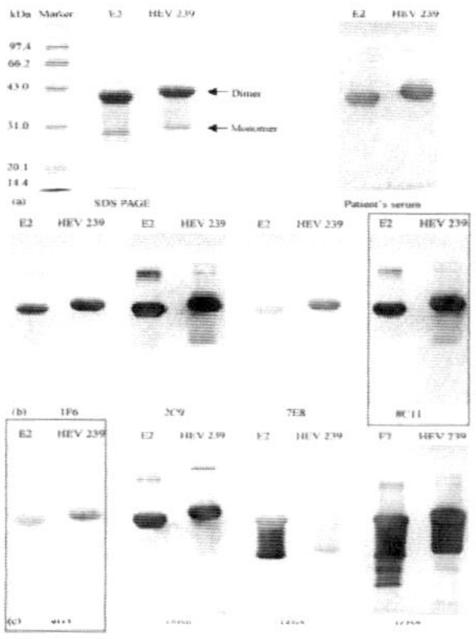


SDS-PAGE and WB of p239 vs. E2

- Successive N terminal extension of E2 generated particulate HEV 239, retaining dimerization capability and major neutralization epitopes.

Li et al. *JBC, 2005; Vaccine.* 2005

National Institute of Diagnosis and Vaccine Development in infectious diseases, Xiamen University

# Enhanced Immunogenicity - ~100-fold enhancement by VLP Ag

Table 1
Antibody response of mice to the HEV 239 and E2 vaccines

| HEV 239 vaccine | | E2 vaccine | |
|---|---|---|---|
| Dose (μg) | Seroconvertion no./ inoculated no. | Dose (μg) | Seroconvertion no./ inoculated no. |
| 20 | 8/8 | 5 | 0/4 |
| 6.67 | 8/8 | 10 | 0/4 |
| 2.22 | 8/8 | 15 | 0/8 |
| 0.74 | 8/8 | 20 | 0/4 |
| 0.25 | 7/8 | 30 | 1/8 |
| 0.08 | 3/8 | 60 | 2/8 |

Balb/c mice were inoculated once with the indicated doses of HEV 239 or E2 and bled 4 weeks later for the determination of HEV antibodies. Both vaccines were suspended in alum adjuvant.

Li et al. Vaccine 2005.

# HEV 239 vaccine efficacy in rhesus monkeys

Virus
dose
(MID50)

3%-99.8%)

Li et al. Vaccine 2005.

National Institute of Diagnosis and Vaccine Development in infectious diseases, Xiamen U

Successful
Scale-up
of p239 VLP
production

HPLC-SEC
DSC

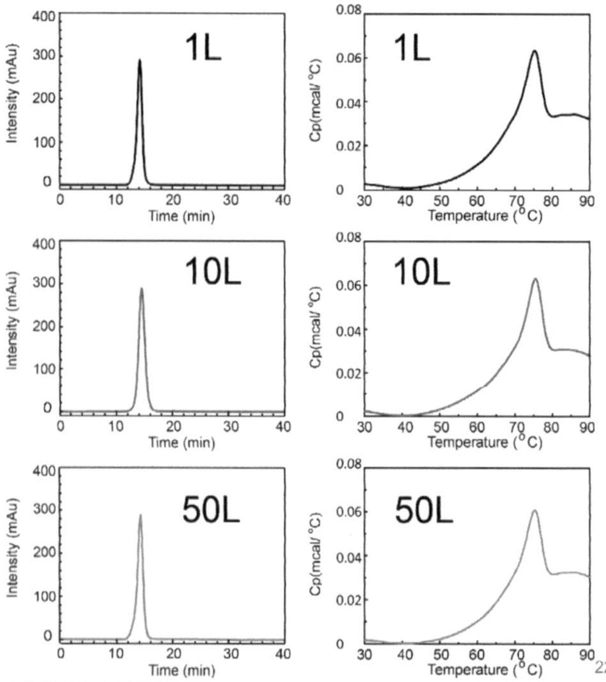


22

National Institute of Diagnosis and Vaccine Development in infectious diseases, Xiamen University

# TEM of p239 VLPs

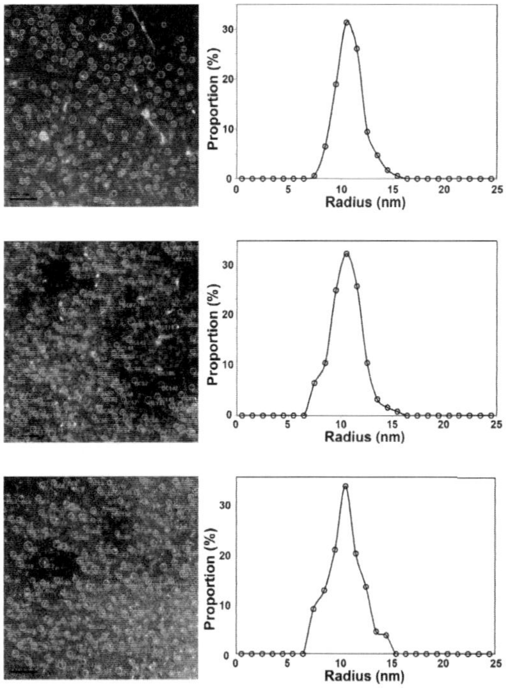


23



# Peptide mapping of 5 lots of p239 VLPs

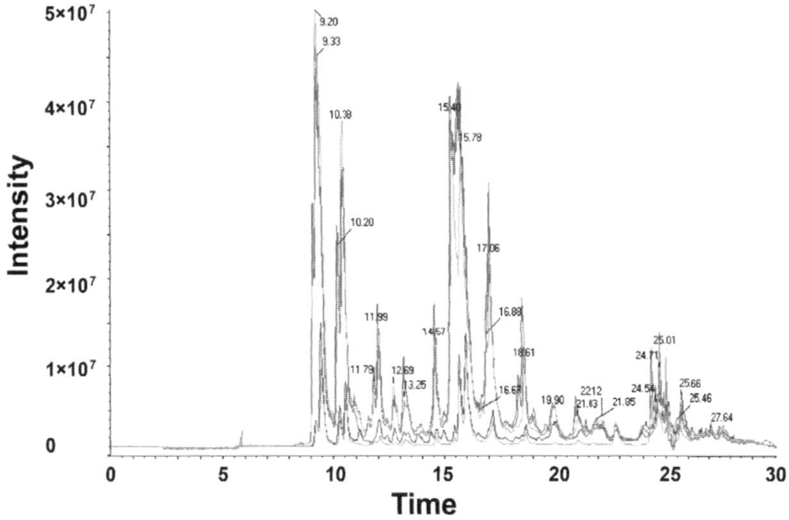


24

# LC-MS-MS mapping of tryptic digested p239 lots

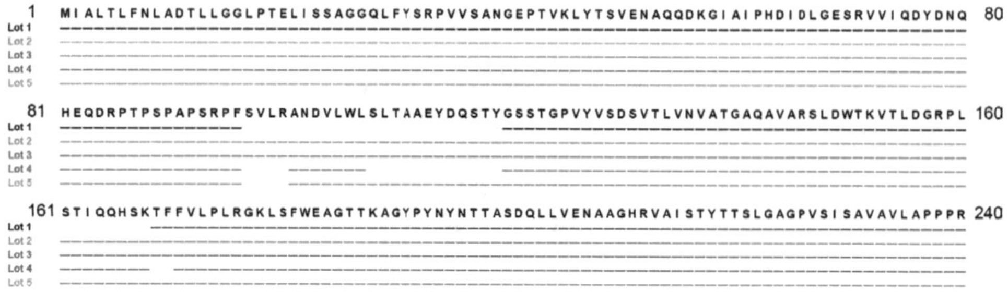


25

National Institute of Diagnosis and Vaccine Development in infectious diseases, Xiamen University

# Inhibition ELISA (IC50) of p239 VLPs with diff. mAbs

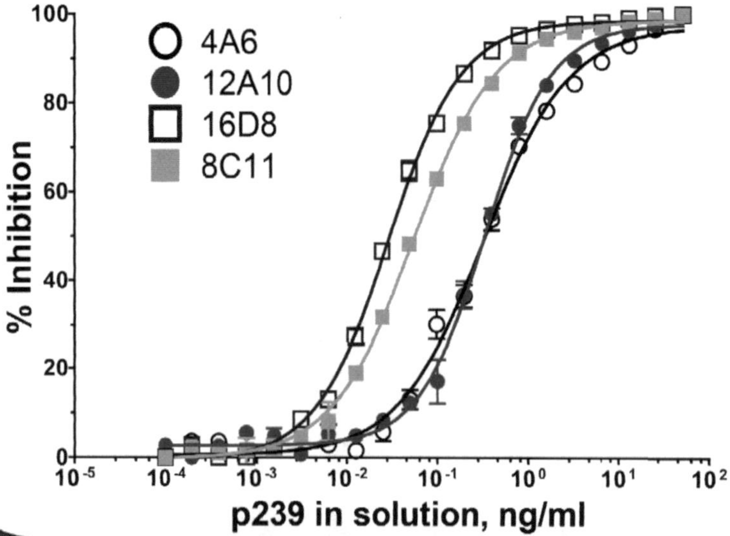


26

National Institute of Diagnosis and Vaccine Development in infectious diseases, Xiamen University

# Epitope multiplicity: same mAb on both sides of a sandwich ELISA

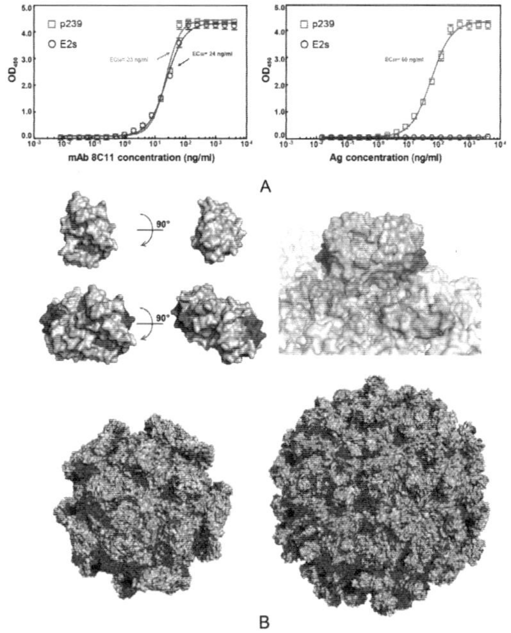




# Multi-epitope VLP probing using a panel of mAbs

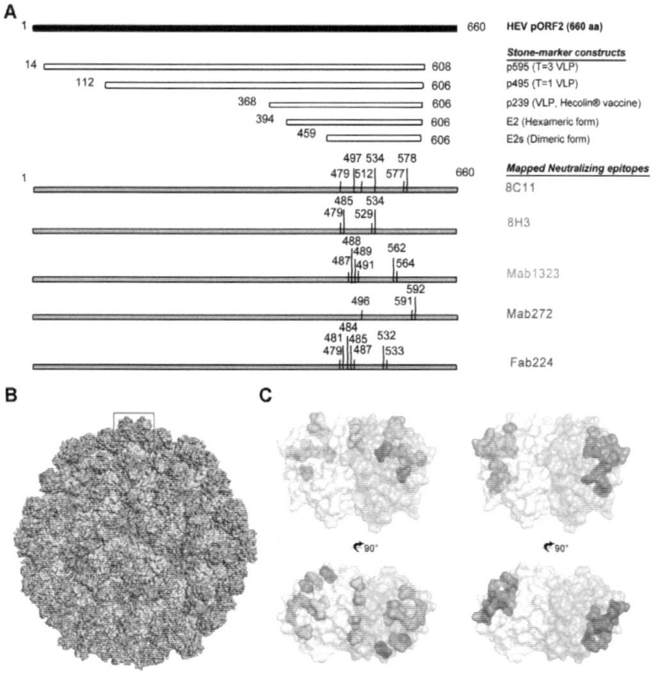

## Multi-angle probing of an elephant!!

29

National Institute of Diagnosis and Vaccine Development in infectious diseases, Xiamen U

## From bench to commercial manufacturing

From bench to clinical (testing)

30

# Clinical trials of HEV vaccine

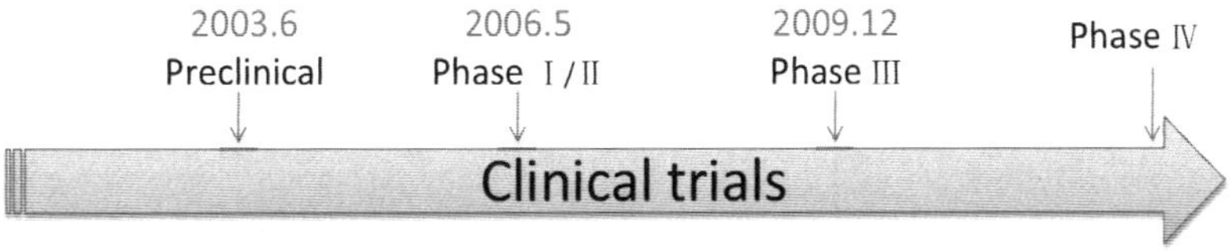

Guangxi, China

Jiangsu, China

National Institute of Diagnosis and Vaccine Development in infectious diseases, Xiamen University

# Clinical trials of HEV vaccine

**Table 1.** Clinical trials of the recombinant HE vaccine during late-stage development of Hecolin®

| Trial | Study Purpose | No. of patients | Key conclusion | Ref. |
|---|---|---|---|---|
| **Phase Ia** (2005.1–2005.3) | Safety | 44 | ✓ Well tolerated* | |
| **Phase IIa** (2005.4–2005.11) | Safety, Dose schedule | 457 | ✓ Well tolerated*<br>✓ 3 doses with a regimen of 0–1-6 min was selected | |
| **Phase IIb** (2005.4–2005.11) | Safety, Dosage escalation | 155 | ✓ Well tolerated*<br>✓ Seroconversion rate was 100% in all the subjects<br>✓ 30-μg dose was selected for a Phase III clinical trial | |
| **Phase III** (2007.8–2009.5) | Safety, Efficacy, Immunogenicity | 112 604 | ✓ Well tolerated in the general population* and showed preliminary evidence for safety in pregnant women.*&<br>✓ Efficacy after three doses was 100% (95% CI 72·1–100).<br>✓ 98.7% of subjects had anti-HEV IgG response after receiving HE vaccine | |

Notes: *No vaccine-related SAEs were reported. [&]68 pregnant women were inadvertently vaccinated with 1 or more doses of HE vaccine or placebo vaccine during Phase III clinical trial; the AEs occurred in this population were analyzed.

Wu, et al. *Human Vaccines*. 2012

# Phase III Clinical trials of HEV vaccine

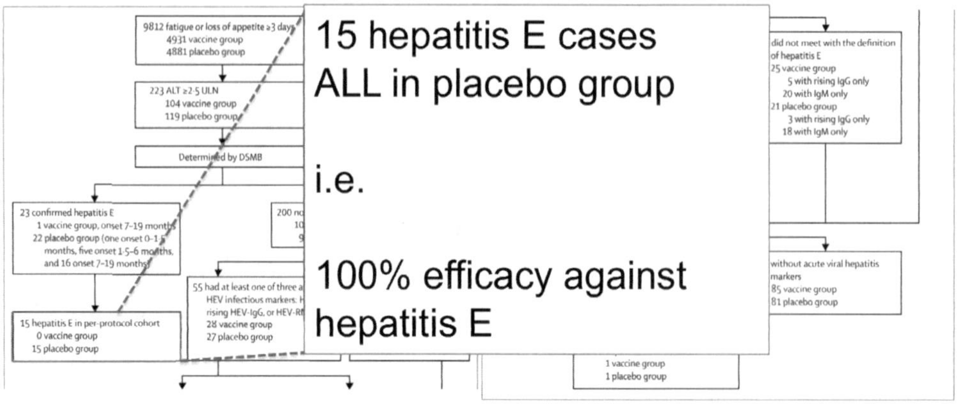


15 hepatitis E cases
ALL in placebo group

i.e.

100% efficacy against hepatitis E

Flow chart of surveillance and certification of acute hepaitis E in phase III

Zhu, et al. *Lancet.* 2010

National Institute of Diagnosis and Vaccine Development in infectious diseases, Xiamen University

# Publications


1. **Wu, et al.** *Hepatology* **2011(in press)**
2. **Tang, et al.** *PNAS* **2011, 108(25):10266-10271**
3. **Zhu F, et al.** *Lancet* **2010, 376(9744):895-902.**
4. **Huang SJ, et al.** *PLoS One.* **2010, 5: e13560**
5. **Yu H, et al.** *J Mol Model* **2010; DOI 10.1007/s00894-010-0794-5**
6. **Guo Q, et al.** *J Clin Microbiol* **2010,48:317-318**
7. **Zheng Z, et al.** *J Gen Virol* **2010,91:1728-1736**
8. **Li S, et al.** *Plos pathogens* **2009, 5:e1000537**
9. **Zhang J, et al.** *Vaccine* **2009,27:1869-1874.**
10. **Wu T, et al.** *Intervirology* **2008, 51:322-317.**
11. **He SZ, et al.** *J Gen Virol* **2008, 89:245-249.**
12. **Chen YW, et al.** *Biomed Environ Sci* **2007, 20:488-494.**
13. **Ge SX, et al.** *Biomed Environ Sci* **2007, 20:521-515.**
14. **Luo WX, et al.** *FEMS. Immunol Med Microbiol* **2007, 51:18–25.**
15. **Wu T, et al.** *Mol Immunol* **2007, 44:3161-3266..**
16. **Zheng YJ, et al.** *J Infect Dis* **2006, 193:1643-1649.**
17. **Li RC, et al.** *Emerg Infect Dis* **2006, 12:1682-1688..**
18. **Li SW, et al.** *J Biol Chem* **2005, 280:3400-3406.**
19. **Zhang J, et al.** *Vaccine* **2005, 23: 2881-2892.**
20. **Li SW, et al.** *Vaccine* **2005, 23: 2893-2901.**
21. **Xia NS, et al.** *Vox Sang* **2004, 86:45-47.**
22. **Zhang J, et al.** *J Med Virol* **2003, 71:518-526.**
23. **Wang YC, et al.** *J Med Virol* **2003, 67:516-521.**
24. **Zhang JZ, et al.** *J Med Virol* **2002, 66:40-48.**
25. **Im S, et al.** *Vaccine* **2001, 19:3726-3732.**
26. **Zhang JZ, et al.** *J Med Virol* **2001, 64:125-132.**

## Other VLP based vaccines under development

- HPV16/18 bivalent vaccine – cervical cancer
- HPV6/11 bivalent vaccine – genital warts
- HPV$_{16/18/52/58/31/33/45/35/59}$ 9-valent vaccine
- RotaV VLP-based vaccine

National Institute of Diagnosis and Vaccine Development in infectious diseases, Xiamen U

## HPV16/18 bivalent vaccine

T=7 HPV16 VLP

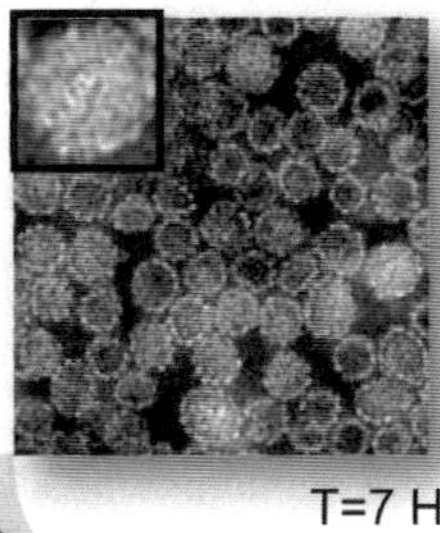

T=7 HPV18 VLP

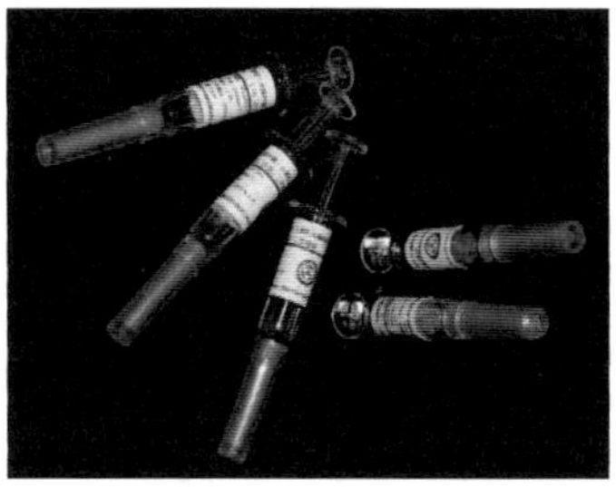

HPV16/18 bivalent vaccine

- *E. coli* expression system
- HPV L1 virus-like particles

National Institute of Diagnosis and Vaccine Development in infectious diseases, Xiamen University

## Immunogenicity of HPV16/18 bivalent vaccine in rehsus monkeys

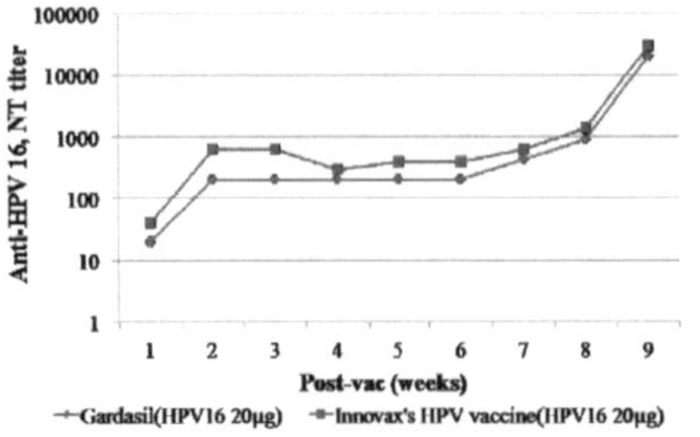


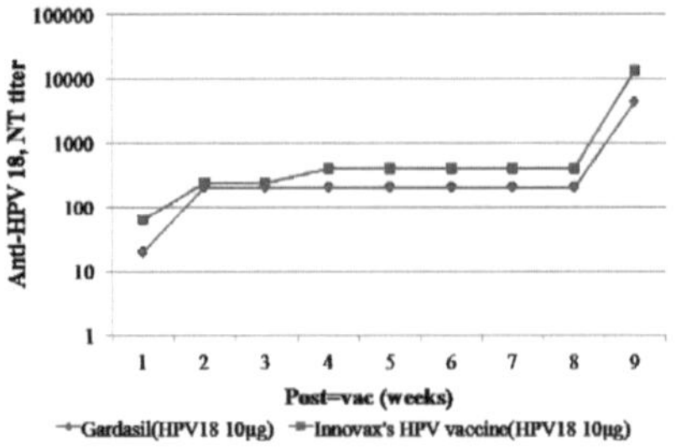


✧ Comparable neutralizing antibody levels

National Institute of Diagnosis and Vaccine Development in infectious diseases, Xiamen University

## HPV6/11 bivalent vaccine

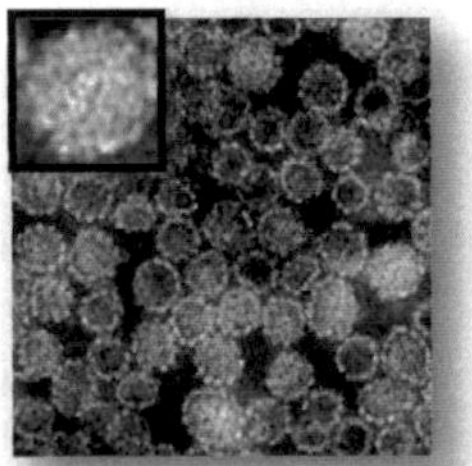

T=7 HPV6 VLP

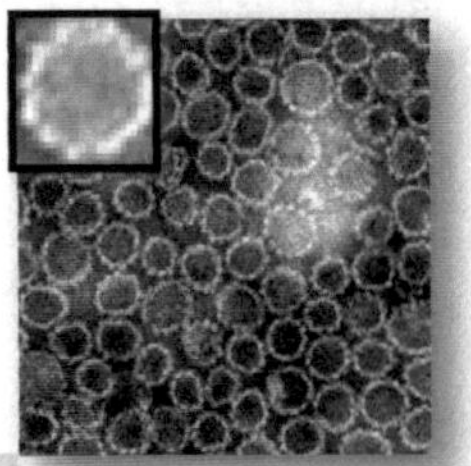

T=7 HPV11 VLP

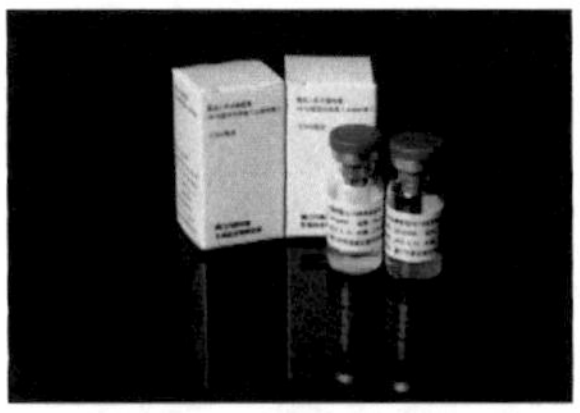

HPV6/11 bivalent vaccine

✧ *E. coli* expression system

✧ HPV L1 virus-like particles

National Institute of Diagnosis and Vaccine Development in infectious diseases, Xiamen University

# On-going efforts on VLP based vaccine

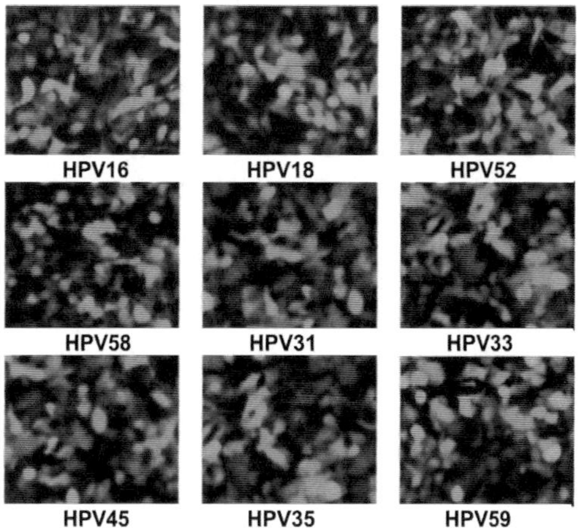


- High-risk type: HPV16/18/52/58/33/31/45/35/59 VLP
- Corresponding pseudovirion-based neutralization assay



# Ongoing efforts on VLP based nano-scale bioparticles

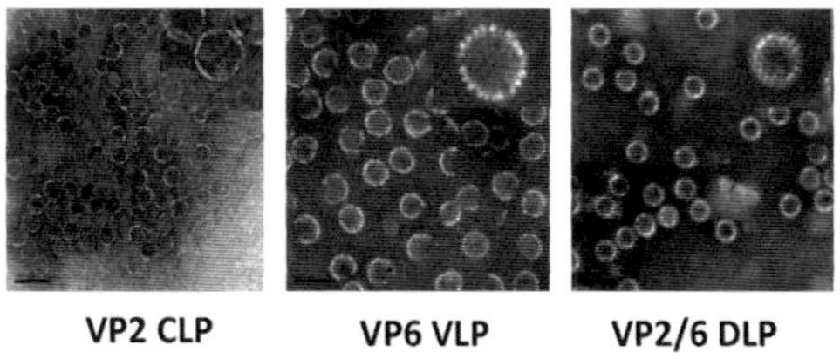


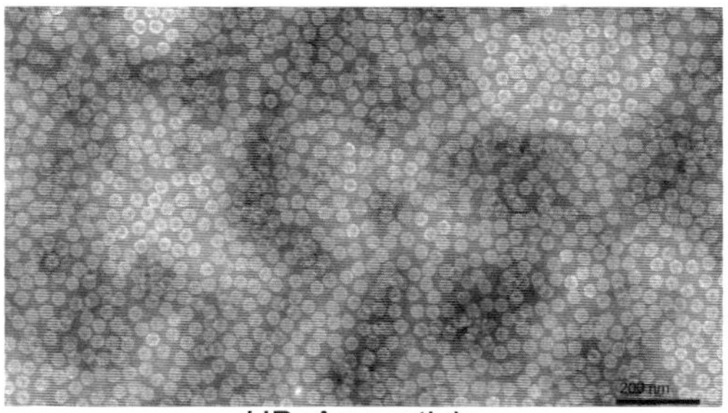

HBcAg particle

- ✧ Rotavirus VLP for vaccine
- ✧ HBcAg for epitope presentation
- ✧ HBcAg used in HB diagnostics

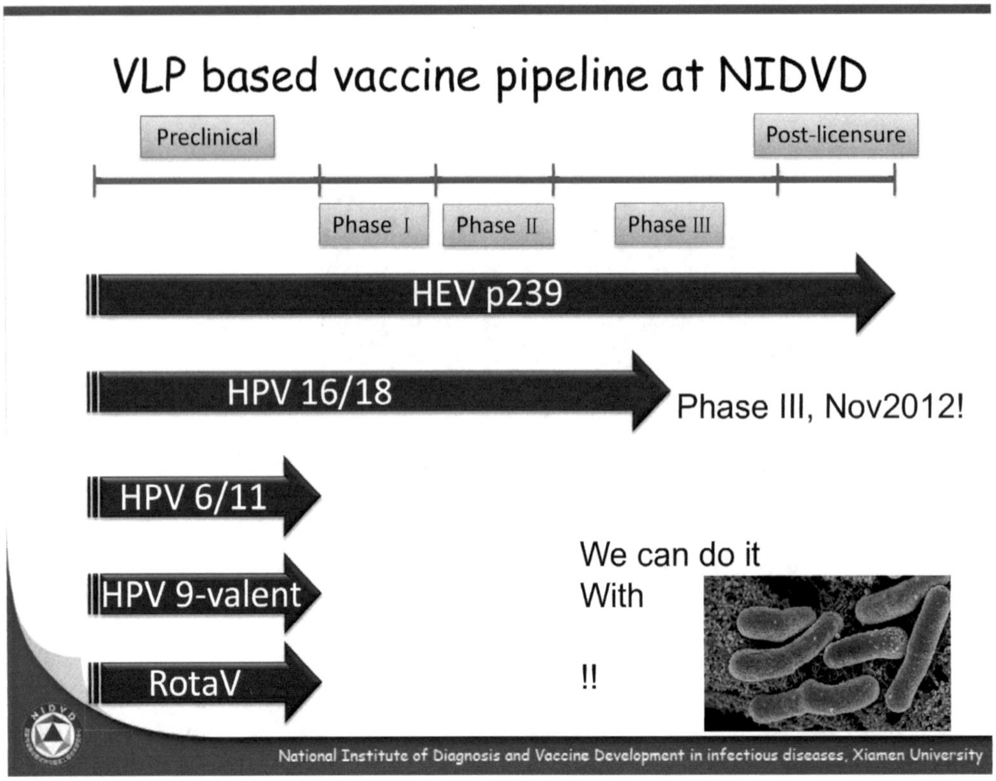


# VLPs as human vaccines?

- ✧ Virion-like or subviral VLPs ("**nature-inspired**")
- ✧ Virion-like epitopes ("**nature-intended**")
- ✧ Epitope multiplicity for effective presentation
- ✧ Stability of vaccines (required for shelf life)

# Toolbox for VLP characterization?

✧ Biochemical (peptide mapping, etc)

✧ Biophysical (size, heterogeneity, aggregation, etc)

✧ Immunochemical (epitopes, affinity to mAbs)

✧ Immunological (mouse potency, GMT, ED50, etc)

National Institute of Diagnosis and Vaccine Development in infectious diseases, Xiamen U

# Acknowledgements


| XMU-NIDVD | NUS | UCSD | Innovax Co. |
|---|---|---|---|
| Shaowei Li | Xuhua Tang | Jack Yan | Huirong Pan |
| Jun Zhang | J Sivaraman | Norm Olson | Yongzhong Gao |
| **Ningshao Xia** | Choyleong Hew | Timothy S. Baker | |
| **And100+ others** | | | |


National Institute of Diagnosis and Vaccine Development in infectious diseases, Xiamen University

# Thank you!

MedImmune

# Development of a Seasonal Vaccine & Strategies for Accelerating Development of Other Bioproducts

Kuldip Sra, Ph.D.

AccBio 2013

Session 4: Analytical Development & Product Characterization Strategies

25-February-2013

## Outline

- Overview
- Seasonal *vs.* Non-Seasonal (Live) Viral Vaccine Development
- A Case Study for a Seasonal (Influenza) Vaccine
- Opportunities to Accelerate Development of Other Biopharmaceuticals?
- See 'A-VAX: Applying QbD to Vaccines' (2012) (http://www.pda.org/Home-Page-Content/CMC-VWG-A-VAX.aspx)

MedImmune

2

## Overview

- In the world of seasonal vaccines you either 'do it fast or fail'
- Focus will be analytical techniques applied to a seasonal vaccine
- Highlight strategies that may be leveraged for accelerating development of other biopharmaceuticals

MedImmune

3

## Development & Manufacturing Challenges for Live Virus Vaccines

- **Common Challenges**
  - Adventitious Agents
  - Stability
  - High Growth with Minimal Process Losses
  - Selection of a Potency Assay
    - Replication / Infectivity (TCID50 *vs.* FFA)
    - Hemagglutination Assay (HA)
- **Unique Challenges Specific to Seasonal Vaccines – one chance to get it right…**
  - Critical Reagents
  - Timeline for Safety Studies
  - Annual Reformulation (Master Seeds)
  - Multivalent Potency Assays

MedImmune

4

## Overview – Cooperative Effort

- ***Extraordinary interaction & cooperation among manufacturers, public health agencies & national regulatory authorities to harmonize seasonal vaccine release...***
- **Key Activities**
  - Control Strategy
  - Critical Assay Reagents
  - Potency – Activity Correlate
  - Selecting the Viruses in the Vaccine
  - Vaccine Supply

MedImmune

5

## Overview – Opportunities to Accelerate Bioproduct Development?

- **Potency correlates to Activity**
- **Line-of-Sight from R&D → Manufacturing**
- **Control Strategy (includes Comparability)**
  - Multifactorial design (DOE) of potency assay & bioprocess
    - Understand & Leverage 'Design Space'
    - Understand impact on CQAs
    - Use mini-blends (pilot scale) to predict optimized blend (at full scale)
  - Leverage Prior Knowledge

A-VAX: Applying QbD to Vaccines
(http://www.pda.org/Home-Page-Content/CMC-VWG-A-VAX.aspx)

MedImmune

6

# Seasonal *vs.* Non-seasonal (Live) Viral Vaccine Development

*For a seasonal vaccine,*

*"if you've seen one influenza season, you've seen one influenza season."*

*— Anonymous*

MedImmune

7

## Vaccines

- **Vaccines stimulate the immune system to protect against or prevent a disease**
  - Relatively complex, difficult-to-characterize biopharmaceuticals
  - Tend to be defined by the 'process' rather than 'properties'
  - Immunity can be humoral (antibody), cellular, or both
- **Types of vaccines include**
  - Inactivated microorganisms (microbial or viral)
  - Live-attenuated microorganisms (microbial or viral)
  - Subunit proteins, peptides, and / or polysaccharides (microbial or viral)
  - Recombinant-derived attenuated or replication incompetent viruses

MedImmune

8

## Seasonal Influenza Vaccines

- In the US, flu season occurs in the fall & winter, peaking anywhere from late November through March
- On average, 5-20% of the population gets the flu & more than 200,000 people are hospitalized from seasonal flu-related complications
- Currently, manufacturers project 145 million doses of flu vaccine to be produced this season (CDC)

MedImmune

9

## 3-D Graphical Representation of a Generic Influenza Virion's Ultrastructure

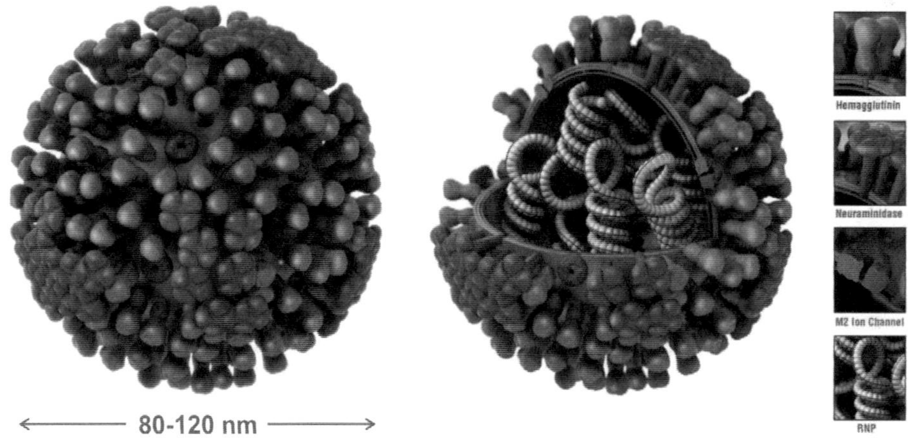


CDC Public Health Image Library

MedImmune

10

## Development & Manufacturing Challenges for Influenza Vaccines

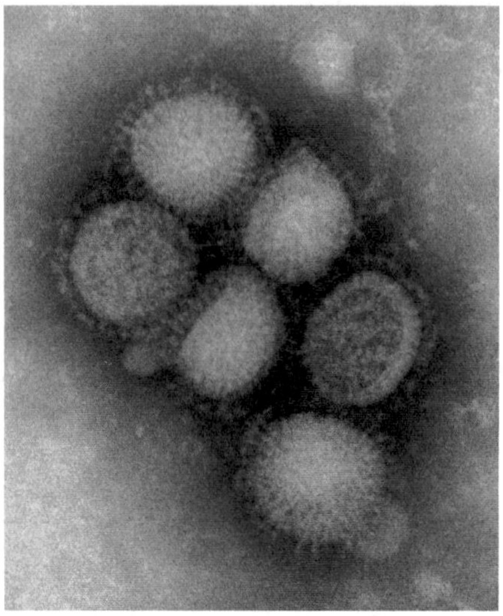

H1N1 influenza virus images from the CDC Influenza Laboratory

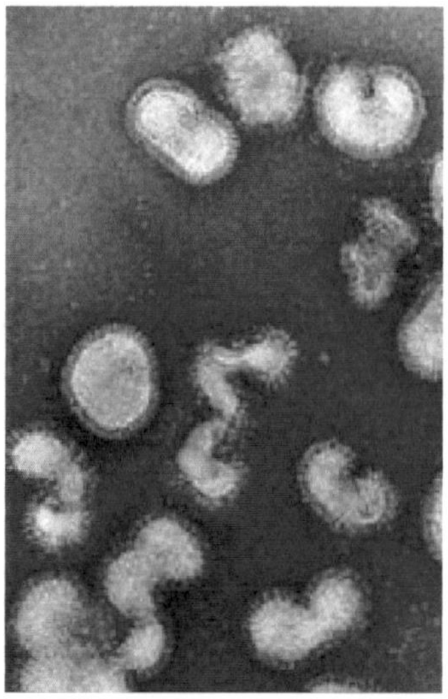

H1N1 (influenza) vaccine images (NanoImaging Services)

MedImmune

11

## Key Differences between Seasonal & Non-Seasonal (Live) Vaccines

| Attribute | Seasonal Vaccine (Influenza) | Non-Seasonal (Live) Vaccine |
|---|---|---|
| API | Seasonal | Unchanged |
| Reliance on Health Authorities (Gating Factor) | Yearly for strain selection | At product license approval |
| Manufacturing Cycle | 5-6 m (parallel process based upon conditional release) | 2-4 y (linear process) |
| Use of Mini-Blends to optimize Blending for multi-valent vaccines | Annually | Once |
| DP Shelf-Life | ≤6 m | >18 m |
| Lot Release | 3 m (conditional release) | >3-6 m (sequential release) |
| Clinical Trials | ~2 m | 2-4 y |

MedImmune

12

## Annual Reformulation (Master Seeds)

- **Most vaccines (Live) are prepared from master seeds that last for the product life span**
  - Manufacturer's working stock seeds may be manufactured multiple times
  - DS bulks may be stored for years
- **Seasonal influenza vaccines are prepared from master seeds that may change annually to match the vaccine strains with contemporaneous circulating virus**
  - New product yearly (requires complete lot-release testing)
  - New master seeds may be manufactured yearly
  - DS bulks may be stored for subsequent year if the strain match is appropriate
  - Blending, filling & labeling of DP is based upon conditional release of DS (sterility & potency)

MedImmune

13

## A compressed, relatively short development time & lifecycle for seasonal vaccines...

**Bioproduct or Non-Seasonal (Live) Vaccine**

**>10 y**

**Seasonal Vaccine**
**<1 y**

MedImmune

14

## Non-Seasonal (Live) Vaccine Development Timeline

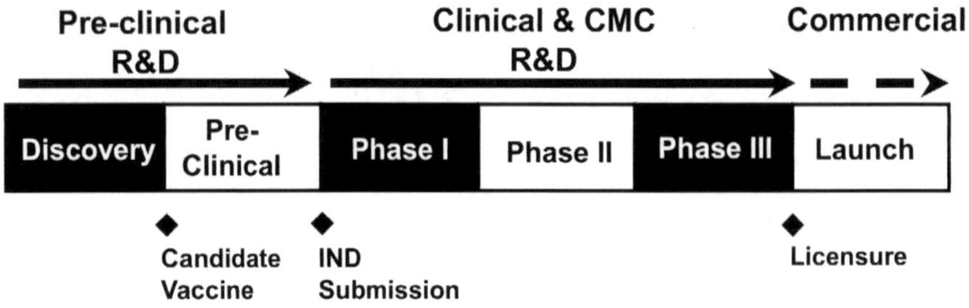


- A relatively long development time & lifecycle…
  - >10 y from discovery to licensure
  - >$800M cost to develop a *prophylactic* product[1,2]
  - Clinical studies are large & complex
  - Product on market for decades

[1] J.A. DiMasi et al: The Price of Innovation. J. of Health Economics, 22 (2003), 151-185.
[2] Source URL: http://www.fiercebiotech.com/story/tufts-billions-wasted-pharma-rd-despite-new-biz-models/2013-01-08

MedImmune

15

## Seasonal (Influenza) Vaccine Development Timeline

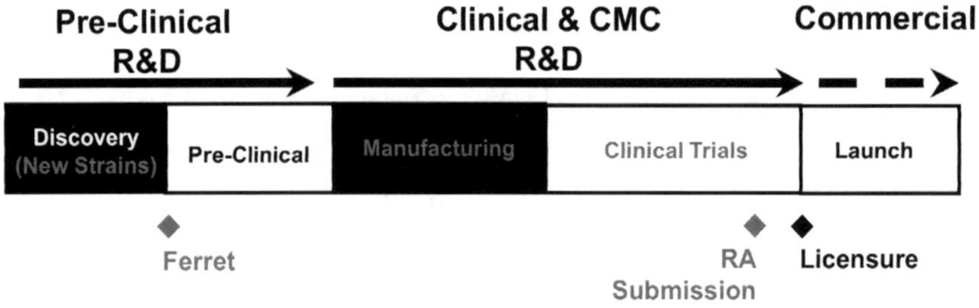


- Each year, it's a new product with a new manufacturing campaign & new stability studies
- A considerable amount of time is spent on steps beyond the manufacturer's control
- A compressed, relatively short development time & lifecycle…
  - 5-6 m from discovery to licensure
  - Abridged Clinical Trials
  - Product on market for one season

MedImmune

16

## Specific Challenges of Producing a Seasonal Vaccine

- **Strain change impacts Annual Safety Trial (AST) which directly affects the launch date** – start time is not flexible
  - AST release is on a short & fixed timeline
  - Conditional release via potency & sterility
  - AST stored at -60 °C
- **Unused formulated & filled vaccine is usually discarded**
- **Monovalent concentrates are not typically used after 12 m from pool**
- **Yields of the strains are variable & are not known by the manufacturer until scale-up production**
- **Adequate supply of MVS bulk?**
- **Fill / Finish Issues?**

MedImmune 17

## Characterization Tests for Seasonal & Non-Seasonal (Live) Vaccines

| Seasonal (Influenza) | Non-Seasonal (Live) | Non-Seasonal (Non-Live) |
|---|---|---|
| • Virus Shape & Size (TEM)<br>• Total Viral Particle Count<br>• Infectious / Total Viral Particle Ratio<br>• HA Content<br>• HA Size & Integrity<br>• Total Protein<br>• Total DNA<br>• Identity (HAI Assay) | • Morphology (TEM)<br>• Total Viral Particle Count<br>• Infectious / Total Viral Particle Ratio | • Physicochemical Properties<br>• Biological Activity<br>• Immunochemical Properties<br>• Purity<br>• Impurity<br>• Contaminants<br>• Quantity<br>• NMR |
| | • Physicochemical Properties<br>• Biological Activity<br>• Immunochemical Properties<br>• Purity<br>• Impurities<br>• Contaminants<br>• Quantity | |

MedImmune 18

## Regulatory Challenges for Release of a Seasonal Vaccine

- **Abbreviated Regulatory-Submission Strategy for a Seasonal Vaccine**
  - Clinical Protocol
  - CTM Release (CoA)
  - IND-CMC Amendment
  - Strain-Change Supplement (includes labeling)
  - Final CSR to Regulatory Agency
  - Lot Release Protocol (LRP)
  - Expedited release of Non-Seasonal (Live) vaccine if the sponsor has licensed seasonal flu vaccine product.
- **Regulatory fillings for** Non-Seasonal (Live) vaccines are a more extended process
- ***Leverage the seasonal vaccine release strategy for release of a Non-Seasonal (Live) vaccine?***

MedImmune

19

## A Case Study for a Seasonal (Influenza) Vaccine

MedImmune

20

Diagram of genetic reassortment in the production of vaccine

- **Conventional Injection & Nasal Spray Influenza Vaccines are Manufactured using Chicken Eggs**
- **Rate-limiting steps…**
  - Selection of Updated Reference Strains
  - Standardizing the antiserum for each strain takes 2 m
  - Availability of strain as an egg-derived isolate

From National Institute of Allergy & Infectious Diseases (NIAID)
Link: File: ReassortmentH1Res.jpg

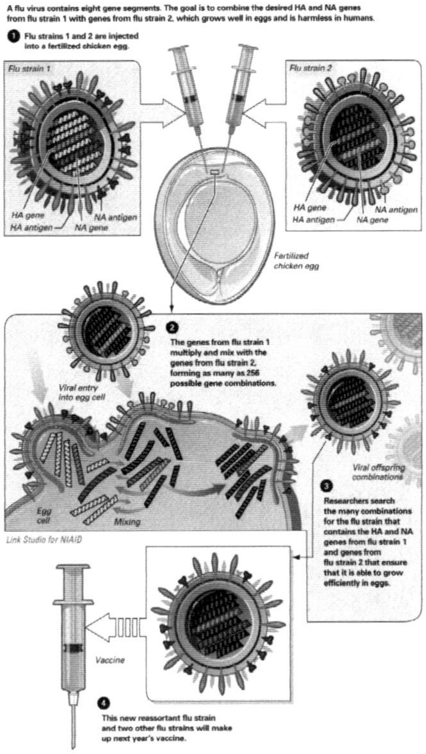


MedImmune

21

## Antigenic Shift – Pandemic Influenza

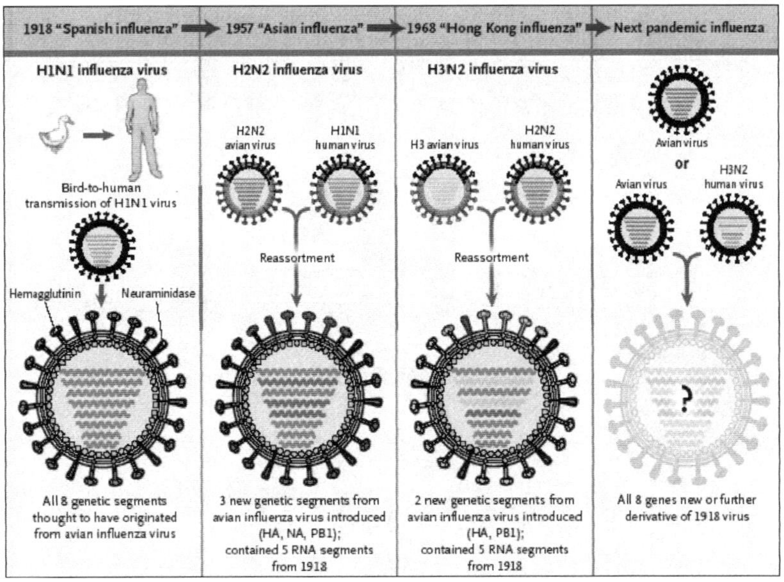


Belshe, R.B. The Origins of Pandemic Influenza – Lessons from the 1918 Virus. *The New England Journal of Medicine* 353:2209-2211, 2005.

MedImmune

22

## Criteria for a Validated Vaccine Candidate

- **High yield in eggs**
  - Rescue variants caused by egg adaptation sequence changes
  - Test growth kinetics & plaque size of different variants
- **Genetically stable**
  - Test HA / NA sequences after egg passages
- **Maintain ca / ts / att phenotypes**
  - ca / ts assay in PCK cells
  - Attenuation test in ferrets
- **Immunogenic in ferrets**
  - 14 day serum HAI / MN against the homologous virus
- **Antigenically accurate (WHO selection)**
  - Serum cross-reactivity to the reference strain & other current strains
- **Manufacturability**
  - EM for virus morphology

MedImmune

23

## Timescale for Broad-Scale Seasonal (Influenza) Vaccine Manufacturing

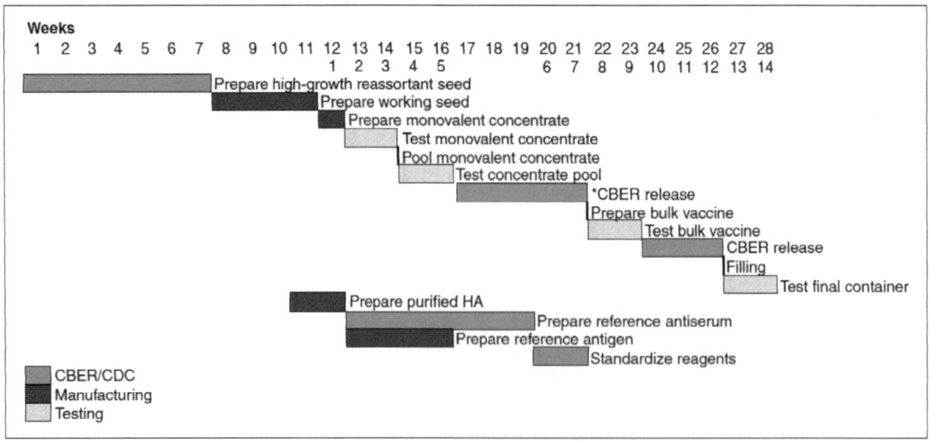


J. T. Matthews (2006) 'Egg-Based Production of Influenza Vaccine: 30 Years of Commercial Experience', http://www.nae.edu/TheBridge, pp. 17-24.

MedImmune

24

## Generalized Timetable for Seasonal (Influenza) Vaccine Production

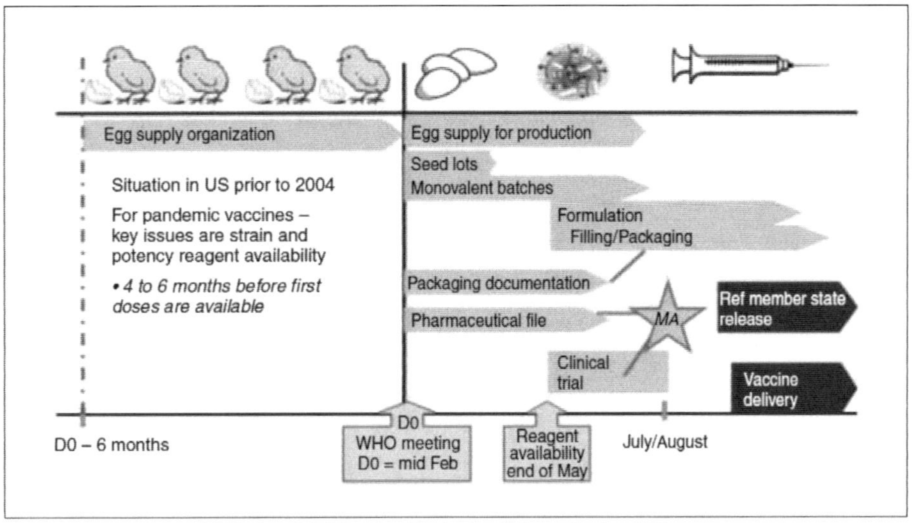


J. T. Matthews (2006) 'Egg-Based Production of Influenza Vaccine: 30 Years of Commercial Experience', http://www.nae.edu/TheBridge, pp. 17-24.

MedImmune 25

## Processes in Seasonal (Influenza) Vaccine Production

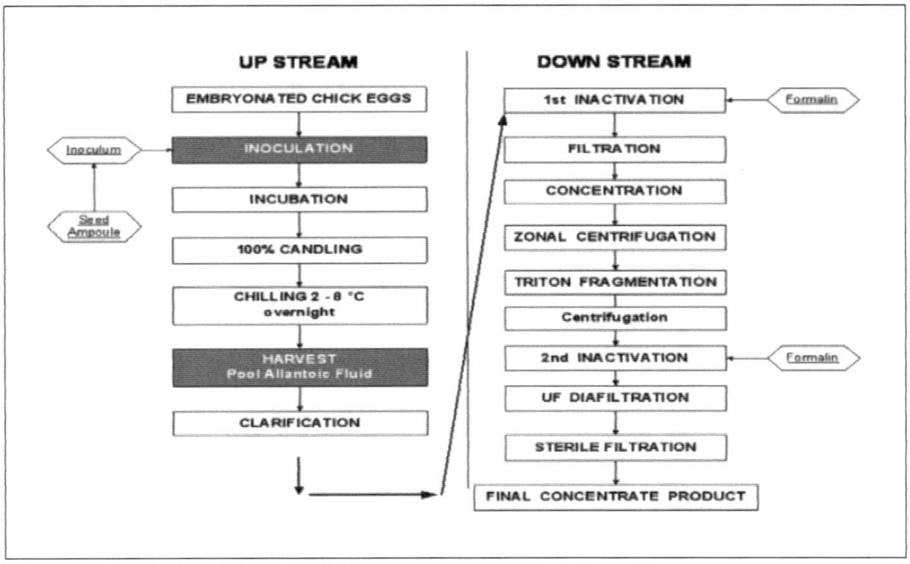


J. T. Matthews (2006) 'Egg-Based Production of Influenza Vaccine: 30 Years of Commercial Experience', http://www.nae.edu/TheBridge, pp. 17-24.

MedImmune 26

## Potency Assays – the Bioassay Continuum

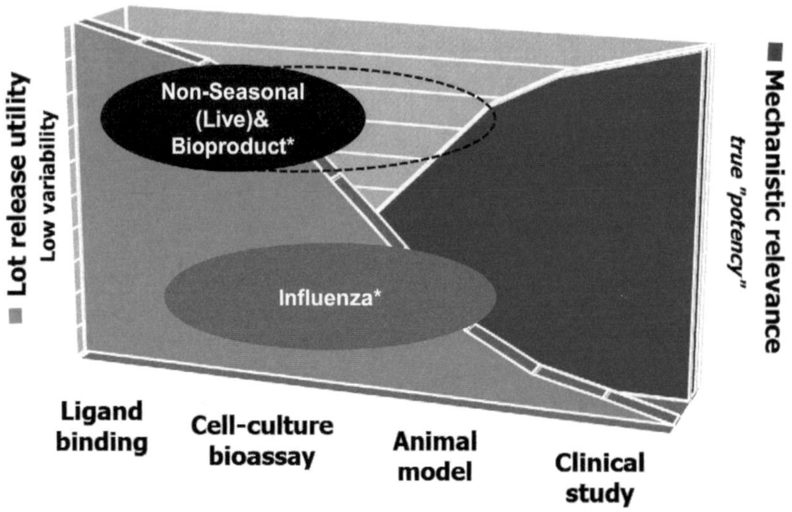


* MedImmune additions to diagram from A. Mire-Sluis & S. Kozlowski

MedImmune

## Annual Safety Trial (AST) Manufacturing & Release Process

| Activities | Seasonal | Non-Seasonal (Live) |
|---|---|---|
| Label Approval | Yearly | • Once during product life-cycle development |
| Quality approves<br>ⓓBatch Record<br>ⓓFinal Product Inspection<br>ⓓLabeling & Packaging<br>ⓓPull Samples (ID & Retains) | Yearly | • Once during product life-cycle development<br>• or when there is a major change in manufacturing process<br>• or during tech-transfer to a different site |
| Quality<br>•MVS & DS bulk<br>•ID & Final Release Testing<br>•DP & Placebo<br>•Packaged CTM<br>•Stability Programs | Yearly | • Once during lot release<br>• Release each lot during product life-cycle<br>• Stability monitoring for >18 m |

MedImmune 28

## Leverage Seasonal Vaccine Knowledge for Non-Seasonal (Live) Vaccines & Other Bioproducts

- **Potency** correlates **to Activity**
- **Line-of-Sight from R&D → Manufacturing**
  - Use shortened development & manufacturing cycle by running the different work streams in parallel
  - Conditional Release
  - Shortened (adaptive) clinical trials to assess safety & Activity
- **Control Strategy (includes Comparability)**
  - Multifactorial design (DoE) of potency assay & bioprocess
    - Understand & Leverage Design Space
    - Understand impact on CQAs
    - Use mini-blends (pilot scale) to predict optimized blend (at full scale)
  - Leverage Prior Knowledge

MedImmune

29

## Acknowledgements

- **MedImmune-Vaccines**
  - Analytical Sciences (Mountain View, CA)
  - Technical Operations (Frederick, MD)
  - Quality Control (Speke, UK)
  - Manufacturing Sciences & Technology (Speke, UK)
- **M. Washabaugh**

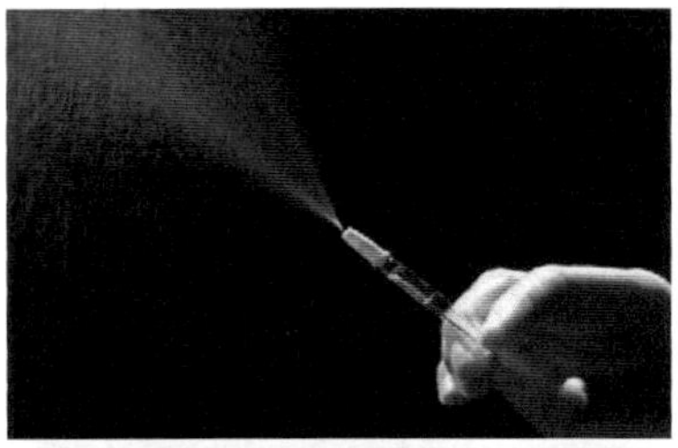

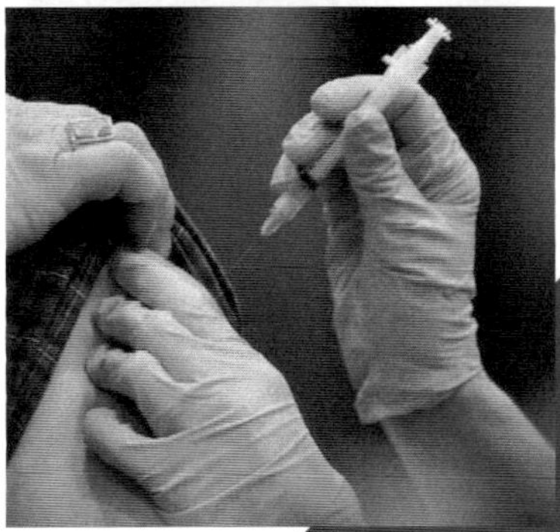

MedImmune

30

**High Throughput Formulation Development of Therapeutic Proteins**

Razinkov, V., Amgen

Originally presented on: 2/26/2013 8:35 - 9:05

Advancements in the fields of molecular biology and biochemistry have led to a significant enrichment in therapeutic proteins pipelines. Currently there are greater than a thousand biotherapeutic products in development and clinical trials. The number of candidates and their variety demands more efficient decision making at both the early and late stage of development. Furthermore, the standard of product quality has improved tremendously throughout the biotech and pharmaceutical industry and it has become extremely critical to accurately assess therapeutic proteins as early as possible. Formulation development is a critical step in the process leading toward efficient, safe and stable commercial products. Due to tight timelines and constrained resources available during early and late formulation studies, the amount of information that can be obtained is often limited. For these reasons, methods with higher throughput capability provide a huge advantage by expanding the conditions and parameters that can be tested. This presentation is focused on the development and application of high throughput methods that can lead to a faster and better evaluation of formulation properties and the improvement of biopharmaceutical development. We describe several analytical and accelerated stress techniques combined into one methodology in order to obtain a comprehensive characterization profile and ultimately to predict quality parameters of protein formulations. Statistical methods of experimental design and analysis were used to estimate the interactions along with significance of multiple factors and to create robustness stability space. Case studies for monoclonal antibody formulations will be presented.

**Utility of High-Throughput Biophysical Analysis of Protein Stability during Formulation Development and Comparability Assessments**

Volkin , D., University of Kansas

Originally presented on: 2/26/2013 9:05 - 9:35

This presentation will examine both opportunities and potential pitfalls of utilizing high-throughput biophysical methods to assess protein stability during formulation development and comparability assessments of therapeutic proteins. The ability of various analytical methods to detect and monitor conformational changes and aggregate formation will be considered though the use of illustrative case studies including the characterization and stabilization of IgG1 and IgG2 mAbs, an albumin-fusion protein, and a recombinant version of a pentameric plasma glycoprotein. Data visualization techniques to analyze and summarize these large biophysical stability data sets will be discussed along with their usefulness for formulation development and potential applicability for comparability studies.

**Application of QbD Principles in Demonstrating Formulation Robustness and Container Closure Compatibility of a Prefilled Syringe Drug Product**

Brader, M., Biogen Idec

Originally presented on: 2/26/2013 9:35 - 10:05

Several case studies of QbD principles applied to formulation robustness of therapeutic monoclonal antibodies (mAb) have been described recently in the open literature and at conferences. These examples have shown elegantly the translation of QbD concepts into practice, but have generally dealt with relatively stable molecules and simple formulations. This presentation will describe a case study of a non-mAb protein susceptible to oxidation and aggregation illustrating how a QbD-inspired approach was applied to demonstrate compositional robustness and container closure compatibility with a new fixed needle prefilled syringe.

# Phase appropriate approach for formulation development

**International Conference on Accelerating Biopharmaceutical Development/ QbD Biologics**

**26th February 2013**

*Robert Mueller,*
*Late-Stage Pharmaceutical and Processing Development,*
*F. Hoffmann-La Roche Ltd., Basel, Switzerland*

## Outline

Phase appropriate approach for formulation development

- **Pharmaceutical Development of Protein Therapeutics**
  - Key challenges
  - Pre-clinical and early clinical formulations
  - Late-stage pharmaceutical development
    - Example: pH/ buffer screen
    - Example: Excipient screen
    - Formulation Robustness
    - Formulation development for combination products
  - Summary

2

## Properties of protein therapeutics

- Large molecules (>10kD)
- Complex molecules
  - 3D-structure which is stabilized by weak non-covalent forces
  - Prone to a wide variety of chemical degradation
- Bioactivity is depend on chemical and structural integrity
- Parenteral administration necessary (if systemic function required)

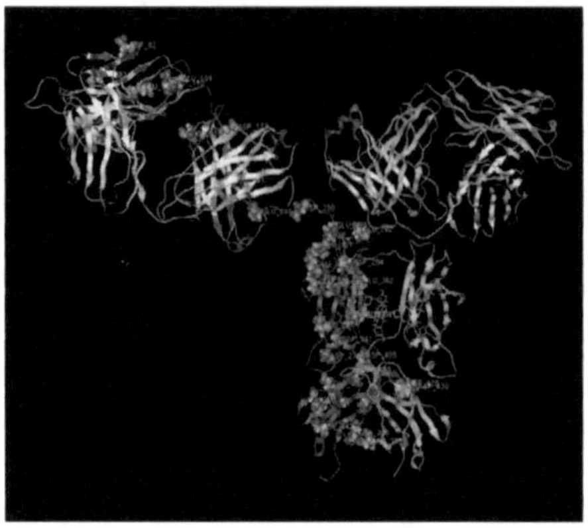

3

## Physical degradation

**Denaturation**

→ Change of native protein structure (e.g change in secondary, tertiary or quarternary structure)

**Adsorption to interfaces**

(air-liquid, ice-liquid, manufacturing materials & primary packaging)

→ Proteins are surface-active!

**Aggregation/Precipitation**

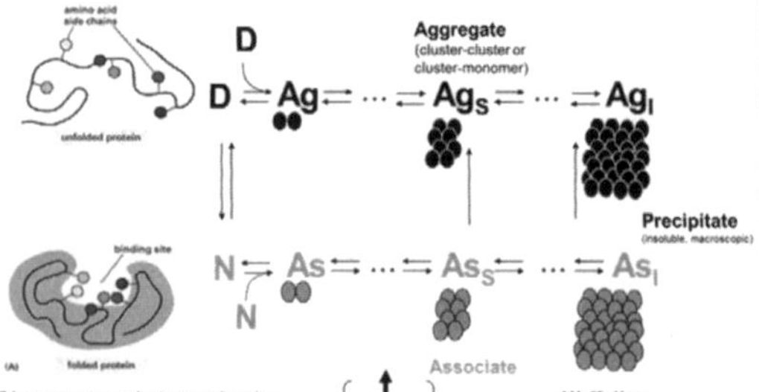


*Other aggregation mechanism, e.g. of covalent nature, could also occur*

* Modified from Brange, 2000; Mahler et al., 2005

4

4

## Chemical degradation

*„by formation/breakage of chemical bonds"*

- Oxidation (e.g. H, M, W,
- Deamidation (e.g. NG)
- Hydrolysis/Proteolysis ( hinge fragmentation)
- Disulfid exchange (C-C
- β-Elimination (C-C)
- Isomerisation (N, D)
- Glycation (K)

• Follow-up reactions like aggregation

• Loss of bioactivity possible

• Influence on PK or bioavail possible

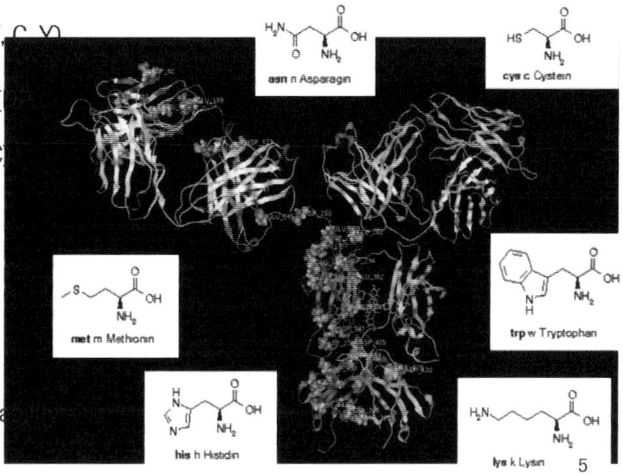


5

## Outline

Phase appropriate approach for formulation development

• **Pharmaceutical Development of Protein Therapeutics**
  - Key challenges
  - Pre-clinical and early clinical formulations
  - Late-stage pharmaceutical development
    - Example: pH/ buffer screen
    - Example: Excipient screen
    - Formulation Robustness
    - Formulation development for combination products
  - Summary

6

## Protein Formulation Strategies

| Tox | Ph1 | Ph2 | Ph3 | Submission, Launch | Line Extension |
|---|---|---|---|---|---|

**Focus of Early-Stage Strategies**

- Ensuring quality (& safety) of the product in preclinical and clinical testing.
- Being sufficiently fast (Timelines to "EI-GLP-Tox, EIH")
- Gated investment
- Considering late-stage development strategies and TPP (!)
  (choice of adequate molecule, „derisk" potential late-stage issues)

**Challenges**

- Availability of (adequate and representative) analytical methods, development material
- Timing of formulation studies and supplies to be closely coordinated with DSP and preclin./clin. studies

## Early Stage Considerations
### *Considering TPP and late-stage strategies*

- Product-related Impurities and
  Formulation Change(s) during Development
  - Clinical exposure
  - Preclinical exposure (tox studies)
- Lyo vs Liquid
  - Stability considerations
- Protein Concentration
- Packaging considerations and Timepoint of Introduction during R&D program
  - Use of Vials for early-stage SC programs
- Route of Administration (i.v.? s.c.?) and Timepoint of Introduction during R&D program

## Early Stage Considerations
### *Early Stage Formulation Strategy Thoughts*

- Which studies should I perform, at which time, and at which concentration?
  1. **Testing of different parameters, such as pH/buffer, excipients, stabilizers, ...**
  2. **„Platform Formulations" for known „Protein classes" (e.g. IgG1s)**

  - At „target protein concentration" (or lower, or higher)?
  - Lyo vs Liquid?
  - Real-time vs accelerated stress conditions?
  - Which analytical methods?

# Early-Stage Form. Dev. Material

- Drug Substance Process Considerations
  - critical for timing of formulation development studies
  - Material from „which process" to use for formulation studies?
  - „representativeness" for later process!?
  - risk of stability failure in case of „non-representativeness"?
- Formulation Development Material „Composition"
  - cannot completely purify from surfactants included in the Development material, use of „flexible material"
- Required amount of protein needed for Form.Dev.
  - depending on the target formulation concentration and amount needed for analytics used in formulation development

## Decrease variability: „standardize & harmonize“

- Applying prior knowledge
- E.g., harmonization and standardization of Drug Product manufacturing processes
  - Use of Disposables (avoiding Cleaning Validation, thus „buying time“) and/or Dedicated Equipment
  - Standard processes, IPCs, Templates
  - Standard freeze-dyring cycles for early-stage use
- E.g., standardization of primary packaging
  - Define some parameters of the final Drug Product to narrow the required Development Space early on
  - Key question: when to change to packaging desired in the TPP?
- E.g. platform formulations
  - To account for the fact that most compounds will fail during further R&D/clinical development

## Outline

Phase appropriate approach for formulation development

- **Pharmaceutical Development of Protein Therapeutics**
  - Key challenges
  - Pre-clinical and early clinical formulations
  - Late-stage pharmaceutical development
    - Example: pH/ buffer screen
    - Example: Excipient screen
    - Formulation Robustness
    - Formulation development for combination products
  - Summary

12

## Protein Formulation Strategies

| Tox | Ph1 | Ph2 | Ph3 | Submission, Launch | Line Extension |
|---|---|---|---|---|---|

### Focus of Late-Stage Strategies

- Ensuring quality (& safety) of the product in clinical testing & the market place.
- Representativeness, comparability
- Process scale-up, facility fit, process robustness & "Quality by design"
- Competetiveness, convenience

### Challenges

- Definition of required clinical dose for market typically only during Phase (2/)3 studies
- „technical frontloading", remaining „risk of failure"

## Strategy for formulation development of protein pharmaceuticals

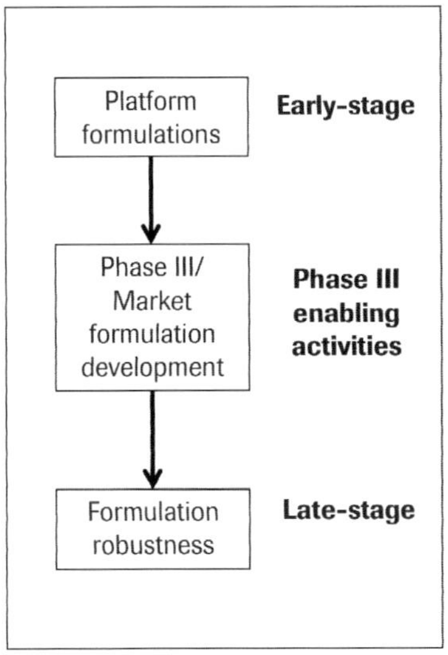


### Phase III/ market formulation development

- Screening studies for
  - pH/ buffer
  - Excipients
- Based on commercial target product profile

### Formulation robustness study

- Market formulations needs
  - to be **stable** (manufacturing, DS & DP storage and administration)
  - and **robus**t (within the limits of critical formulation parameters)

14

## Outline

Phase appropriate approach for formulation development

- **Pharmaceutical Development of Protein Therapeutics**
  - Key challenges
  - Pre-clinical and early clinical formulations
  - Late-stage pharmaceutical development
    - Example: pH/ buffer screen
    - Example: Excipient screen
    - Formulation Robustness
    - Formulation development for combination products
  - Summary

15

## Formulation Robustness study

**DoE/ Experimental Design**

- Full or fractional factorial designs
- Example:
  - $2^4$ experiments + 3 repetitive centerpoints (19 formulations)
  - Interactions covered by the DoE setup (multivariate)

**Factor ranges setting:**

- Ranges to be tested for formulation parameters dependent on
  - prior knowledge
  - project specific specification settings
  - allowable manufacturing ranges etc.

| Formul. | Protein conc. (mg/mL) | pH | Surf. (% w/v) | Stabilizer (mM) |
|---|---|---|---|---|
| F1 | - | - | - | - |
| F2 | + | - | - | - |
| F3 | - | + | - | - |
| F4 | + | + | - | - |
| F5 | - | - | + | - |
| F6 | + | - | + | - |
| F7 | - | + | + | - |
| F8 | + | + | + | - |
| F9 | - | - | - | + |
| F10 | + | - | - | + |
| F11 | - | + | - | + |
| F12 | + | + | - | + |
| F13 | - | - | + | + |
| F14 | + | - | + | + |
| F15 | - | + | + | + |
| F16 | + | + | + | + |
| F17 (target) | 0 | 0 | 0 | 0 |
| F18 (target) | 0 | 0 | 0 | 0 |
| F19 (target) | 0 | 0 | 0 | 0 |

## Formulation Robustness Data Analysis

*Use of Regression Model - Contour Plots*

**Example Contour Plots for HMW Species (3 month 40°C):**
Visualization of predicted response values (spanned by two factors)

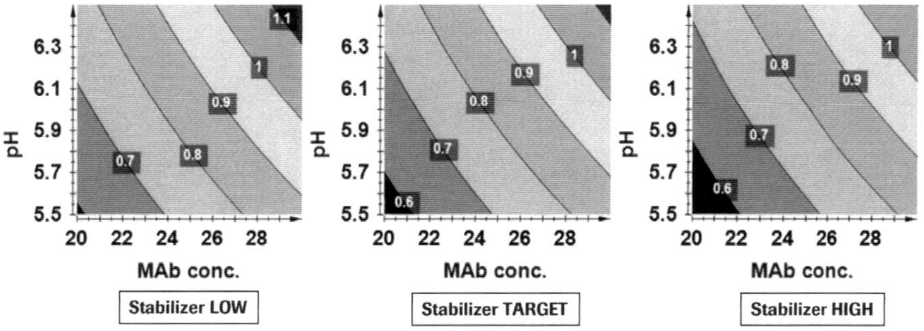


**A formulation** can be claimed as **«robust» if critical quality attributes** (CQAs) **remain within their CQA Acceptance Criteria** (CQA AC) **until end of shelf life within the entire claimed formulation composition design space**.

17

## Outline

Phase appropriate approach for formulation development

- **Pharmaceutical Development of Protein Therapeutics**
  - Key challenges
  - Pre-clinical and early clinical formulations
  - Late-stage pharmaceutical development
    - Example: pH/ buffer screen
    - Example: Excipient screen
    - Formulation Robustness
    - Formulation development for combination products
  - Summary

18

## Formulation Development for combination products

Roche

### Chemical compatibility

- Extractable profile
- Leachable studies
- Tungsten spiking
- Silicone interaction
- Sub-visible particle characterization

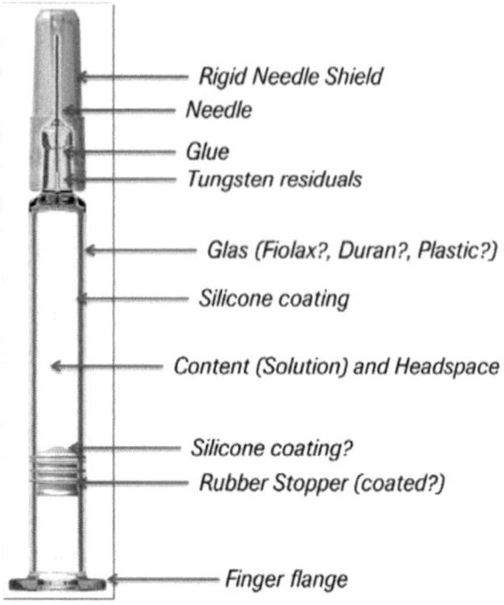


19

## Combination product – high concentration protein formulation

Roche

Impact of single components on performance of Combination Product

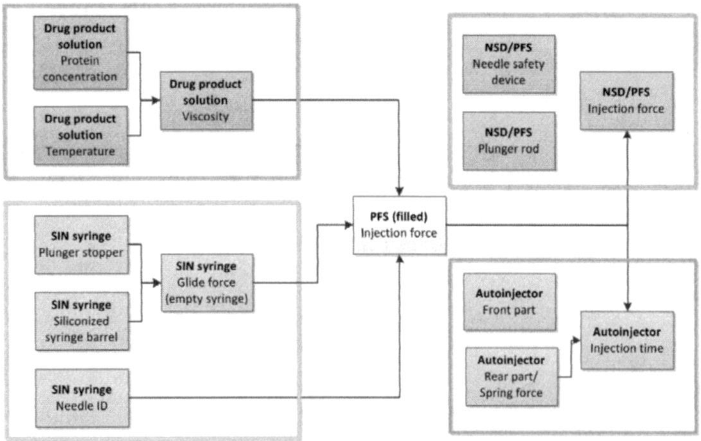

## Summary

- An early stage protein formulation should ...
  - always fulfill safety considerations
  - be very quick
  - build on knowledge!
  - consider the intended use (TPP) and late-stage strategies
  - apply adequate stress and analytical methods
  - generate real-time data as well
- A late stage protein formulation should ...
  - be based on thorough understanding of the physical and chemical properties of a protein
  - demonstrate robustness within the formulation design space
  - consider compatibility with primary packaging and device technology

21

## Acknowledgement

Hanns-Christian Mahler
Michael Adler
Stefan Fischer
Ulla Grauschopf
Pierre Goldbach
Christine Wurth

Karin Schönhammer
Mark Chipperfield
Florian Wildenhahn
Sonoko Kanai
Patrick Bulau
Bernd Moritz

22

Roche

*We Innovate Healthcare*

23

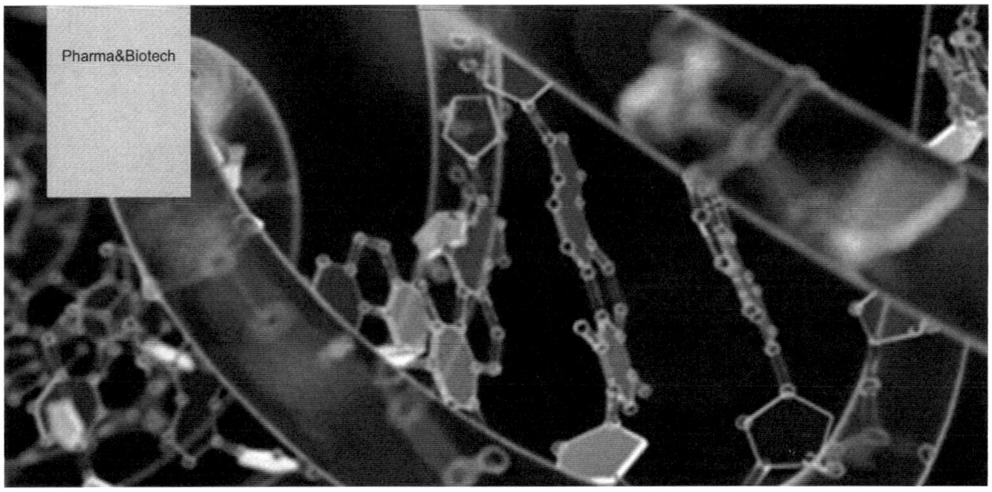


# QbD across the Development Lifecycle: the CMO perspective

SBE 3rd International Conference on Accelerating Biopharmaceutical Development, Feb 26th 2013.

Lonza

John Crowley
Head Cell Culture Process Development

Lonza

## Disclaimer

- Certain matters discussed in this presentation may constitute forward-looking statements. These statements are based on current expectations and estimates of Lonza Group Ltd, although Lonza Group Ltd can give no assurance that these expectations and estimates will be achieved.
- The actual results may differ materially in the future from the forward-looking statements included in this presentation due to various factors. Furthermore, Lonza Group Ltd has no obligation to update the statements contained in this presentation.

*Note: All slides are incomplete without verbal comments.*

Lonza

## Content Overview

1. How it gets done?
- Technical advancements with Process Development workflow – “Islands of Automation”.
  -Linking “Islands of Automation”.

2. What gets done?
- Changed Regulatory Background
- QbD across Development Lifecycle
  - Phase 1: Data gathering
  - Phase 2: Data Interpretation & demonstration
  - Phase 3: BLA submission
- CMO perspective – Lessons learned
- Acknowledgements

3
Mrz-13

PharmaBiotech

## Technical advancements with Process Development Workflow - “Islands of Automation”

Lonza

Lonza

## Key Activities to Release of DS / DP*: Process Development is on the Critical Path

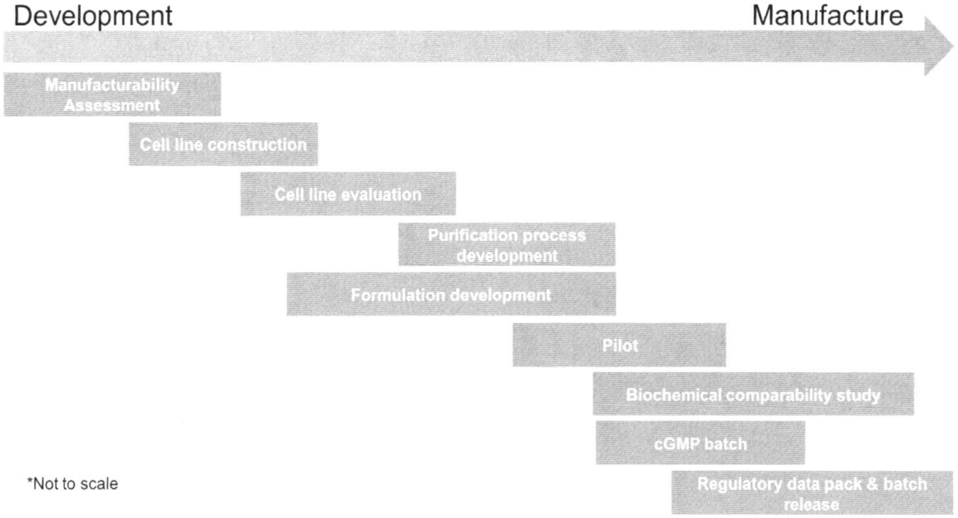


*Not to scale

5
Mrz-13

Lonza

## Process Development Activities are not Always Linear!

A flexible workflow is required to address changing customer demands:

- Transfer of existing cell lines, processes, and assays
- Troubleshooting existing processes
- Addition of work packages

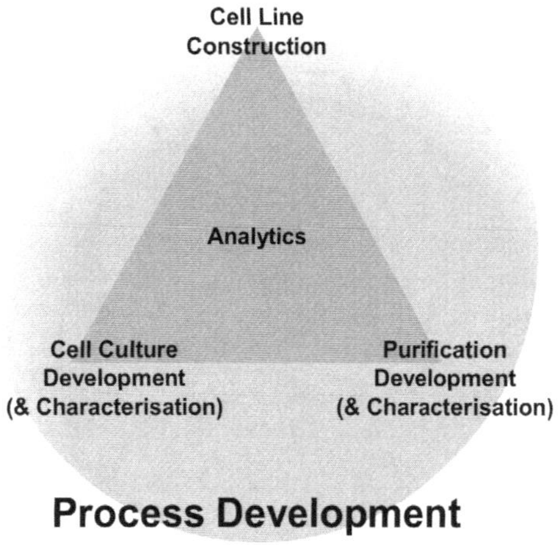


6
Mrz-13

## Lonza's Approach to debottleneck the Process Development workflow

Lonza

- Create "**islands of automation**" to address bottlenecks in the process development workflow
  - Allows flexibility in joining processes together
  - Maintains project segregation

- Automation systems work synergistically to form an "archipelago of automation"
  - Increase sample throughput
  - Supports QbD initiatives
  - Reduction in timelines

7
Mrz-13

## Elements of "Islands of Automation" - Process Development Workflow

Lonza

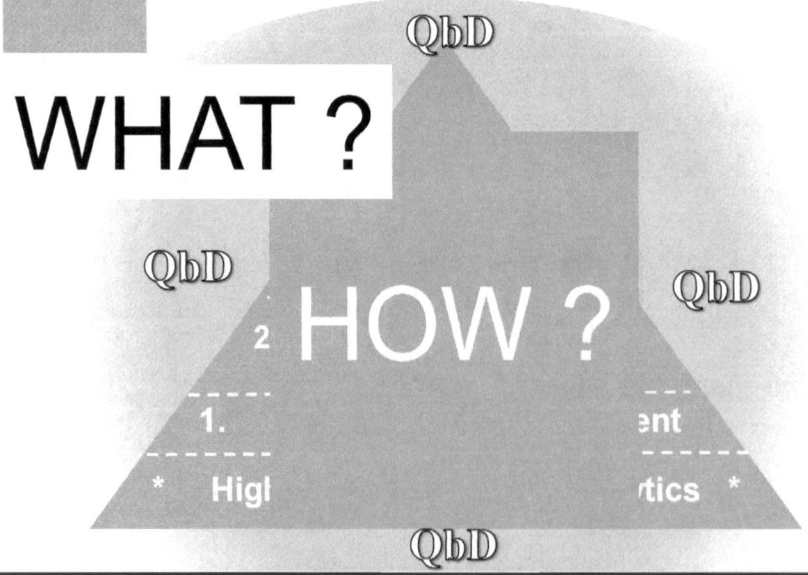


**Development of Cell Culture and Purification Process as Single Entity**

PharmaBiotech

# HOW ?

## Technical advancements with Process Development Workflow - "Linking Islands of Automation"

## Linking Islands of Automation:
## 1. High Throughput Cell Line Development

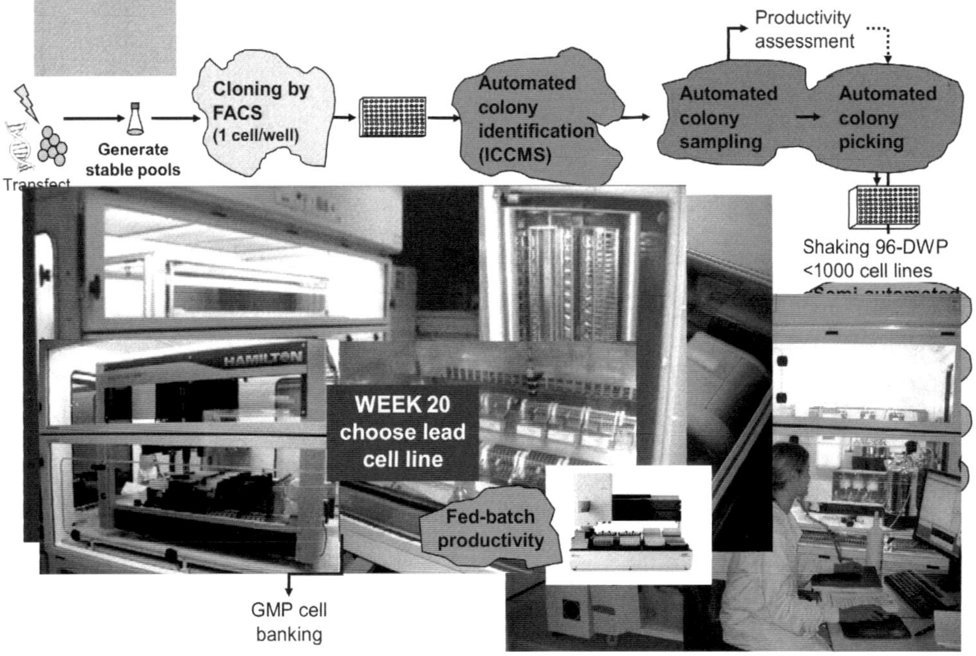

## Linking Islands of Automation: 2. High Throughput Cell Culture Development

Lonza

**Case Study: Comparability of Culture Growth and Viability from 20,000 L to 15 mL**

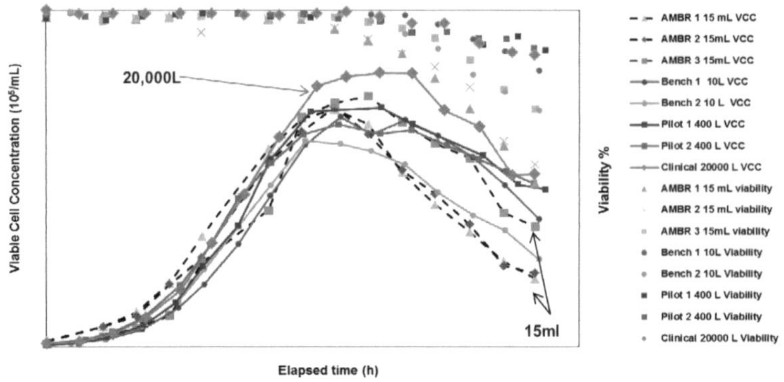


- Comparable growth profiles to the 10 L bioreactor
- Increased understanding of process control and operating parameters essential to more closely mimic performance at scale

11
Mrz-13

## Linking Islands of Automation: 3. High Throughput Purification Development

Lonza

Tecan Freedom Evo 200

- 5 mm internal diameter and 30 mm bedheight
- Designed for resin screening and method development
- Mimic the sequence of real chromatographic steps
- Fractions can be collected in microtiter plate below or discarded

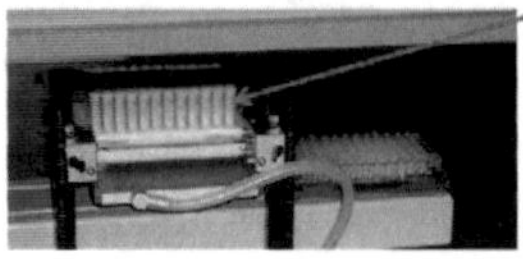

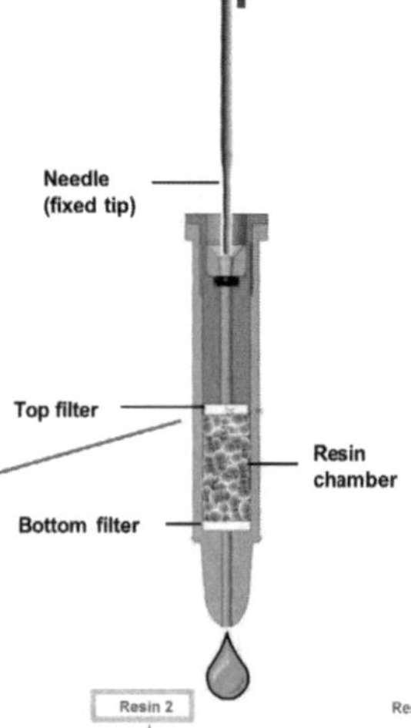


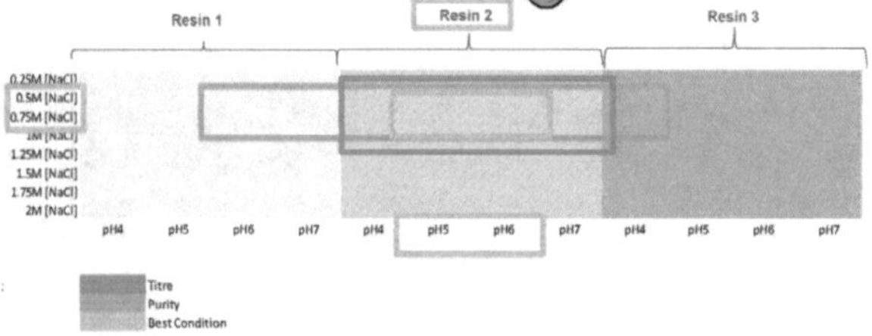

# Linking Islands of Automation:
## * High Throughput Analytics Platform *

**Lonza**

Level I — Level II — Level III

| Level I | Level II | Level III |
|---|---|---|
| CCS Purification – PhyTips<br>SDS, Glycan Screening<br>Protein Concentration (UV) | Sample Preparation, IEF capability<br>Increased throughput (In planning) | Aggregation, full range Glycan, Sialic Acid<br>Kinetics / Activity analysis (In planning) |
| **Sample Prep** | **Fully Automated Systems** | |

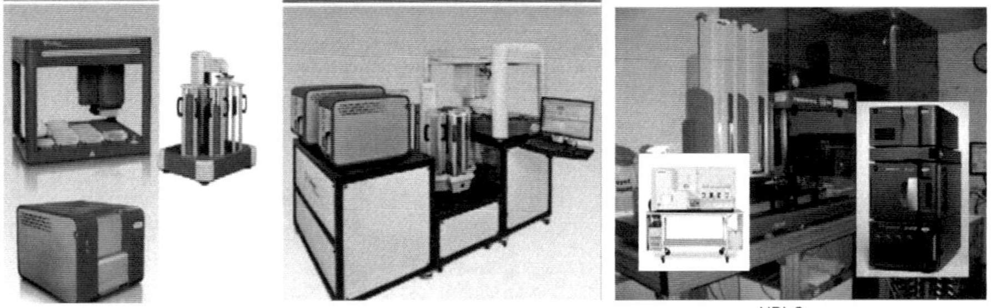

Octet - Caliper GXII/Sciclone UPLC

13
Mrz-13

# Case Study:
## Rapid Product Quality

**Lonza**

**SDS PAGE Traditional Method**

1.5 days

| 15 min | 2 to 3 hours | 2 to 3 hours | 1 hour |
|---|---|---|---|
| Prep Gel | Run Gel | Stain/Destain | Scan/Analyze |

| 15 min | 25 min | X 10 sample capacity |
|---|---|---|
| Prep Chip | Run Chip | Analysis |

2 hours

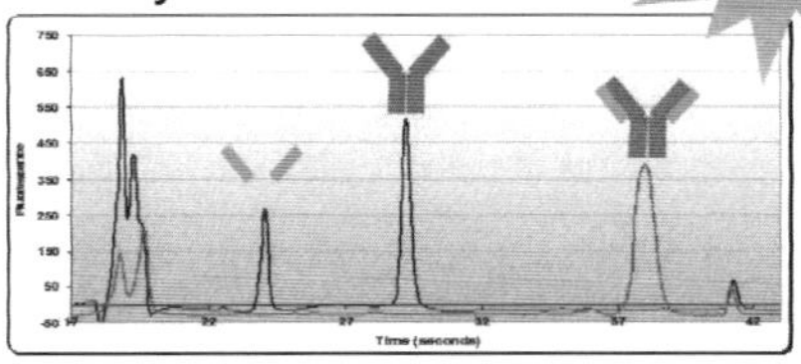

14
Mrz-13

## Elements of "Islands of Automation" - Process Development Workflow

Lonza

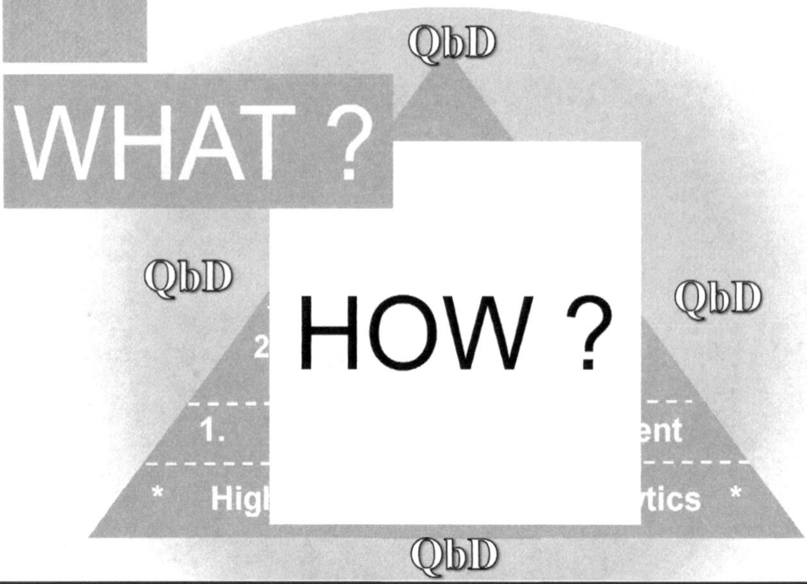


Development of Cell Culture and Purification Process as Single Entity

PharmaBiotech

WHAT ?

**Changed Regulatory Environment**

Lonza

## Changed upfront definition of Process Validation

Lonza

(Echoing throughout all stages of Process Development)

- "Establishing documented evidence that a specific process will consistently produce a product meeting its predetermined specifications and quality attributes."

FDA, May 1987.

Ref: ICH Q7

- "the collection and evaluation of data from the process design stage throughout production, which establishes *scientific evidence* that a process is capable of consistently delivering quality products"

FDA, Nov 2008.

Ref: ICH Q8/9/10/11

## Changed Environment – QbD Approach

Lonza

- QbD: **Quality by Design** - "A systematic approach to development that begins with predefined objectives and emphasizes product and process understanding and process control, based on sound science and quality risk management."

Ref: ICH Q8/9/10/11

PharmaBiotech

# WHAT ?

## QbD across Development Lifecycle
## Phase 1 – Data Gathering

## Scientific based quality risk management:
## Phase 1: Data Gathering

1. Risk Identification → 2. Risk Analysis → 3. Risk Evaluation

**1. Risk Identification**
- ✓ Prior knowledge
- ✓ Process maps
- ✓ Cause & Effect matrix

**2. Risk Analysis**
- ✓FMEA – equipment, cause & effect
- ✓ Finalisation of exp. plan

**3. Risk Evaluation**
- ✓ DOE/OFAT expt.
- ✓ Design space determination

Use of QRM toolbox

Appropriate SME input from all impacted parties critical.

Experimentation

As per ICH Q9

Lonza

## Quality Risk Management Toolbox

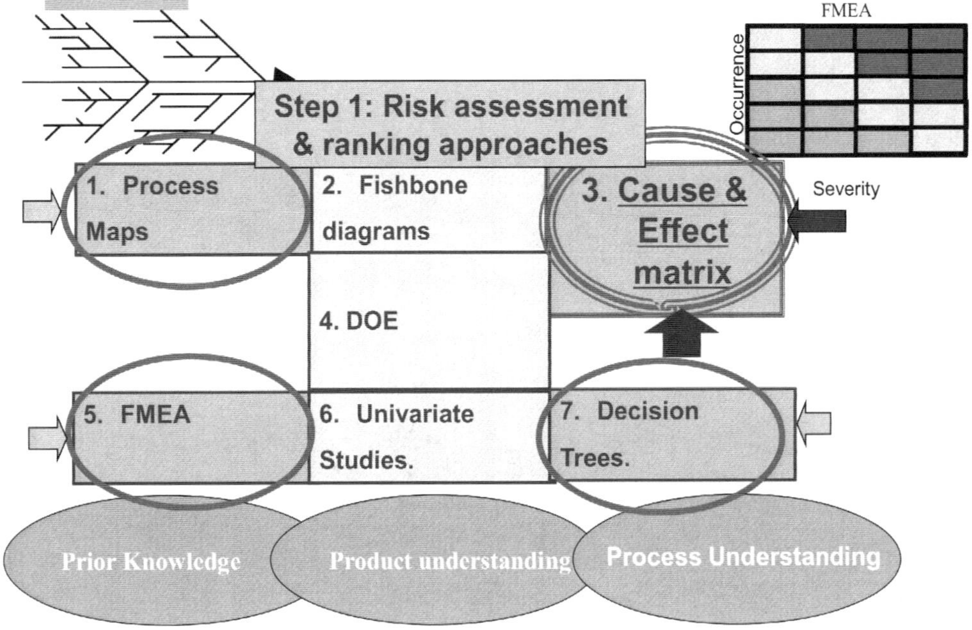


Lonza

## Scientific based Quality Risk Assessment Evolution

### Classic FMEA to detailed <u>Cause & Effect Matrix</u>

- ✓ The Cause and Effect matrix is a detailed screening risk assessment tool used to decide which process inputs require wet work and which do not. The Cause and effect matrix uses all prior product specific data in combination with the team's knowledge to weigh each input according to their impact on specific product quality or performance features.

- ✓ This is an iterative document during a product/process lifecycle which is the overall arching bible used for BLA submission with detailed rationale on parameter designation at its final version.

- ✓ Lonza based cause and effect template used for all unit operation steps from vial thaw to viral filtration for all sites of operation – UK, USA, Singapore and Spain i.e. 200L – 20,000L scales.

## Cause & Effect Matrix

**Lonza**

Production bioreactor step example

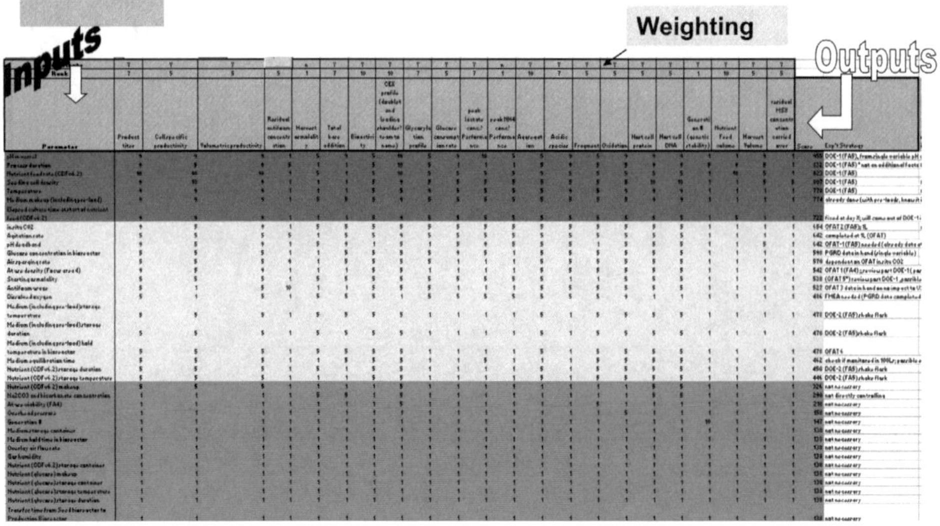


**Risk Identification: "Systemic use of information referring to the risk question or problem description."**

**As per ICH Q9**

## Cause & Effect Matrix

**Lonza**

Chromatography step example

| OUTPUTS / INPUTS | Step yield | Elution Volume | Protein Concentration | Processing Time | DNA | HCP | PrA | intact Ab (nr SDS) | H+L chains (r SDS) | aggregation | fragmentation | acidic variants | antifoam | MSX | oxidation | insulin | Total Score |
|---|---|---|---|---|---|---|---|---|---|---|---|---|---|---|---|---|---|
| Elution Buffer pH | 35 | 25 | 15 | 15 | 10 | 10 | 50 | 50 | 50 | 50 | 50 | 50 | 5 | 5 | 50 | 5 | 475 |
| Load flow rate | 35 | 25 | 15 | 27 | 50 | 50 | 50 | 10 | 10 | 50 | 50 | 10 | 25 | 25 | 10 | 25 | 467 |
| Loading capacity | 63 | 25 | 15 | 27 | 50 | 50 | 10 | 10 | 10 | 10 | 10 | 10 | 25 | 25 | 10 | 25 | 375 |
| PLW1 flow rate | 7 | 5 | 3 | 15 | 50 | 50 | 50 | 10 | 10 | 10 | 10 | 10 | 25 | 25 | 10 | 25 | 315 |
| PLW2 volume | 7 | 5 | 3 | 15 | 50 | 50 | 50 | 10 | 10 | 10 | 10 | 10 | 25 | 25 | 10 | 25 | 315 |
| PLW2 flow rate | 7 | 5 | 3 | 15 | 50 | 50 | 50 | 10 | 10 | 10 | 10 | 10 | 25 | 25 | 10 | 25 | 315 |
| Elution Buffer flow rate | 7 | 25 | 15 | 15 | 50 | 50 | 10 | 10 | 10 | 10 | 10 | 10 | 25 | 25 | 10 | 25 | 307 |
| PLW1 volume | 7 | 5 | 3 | 3 | 50 | 50 | 50 | 10 | 10 | 10 | 10 | 10 | 25 | 25 | 10 | 25 | 303 |
| Load concentration | 35 | 5 | 15 | 15 | 10 | 10 | 10 | 10 | 10 | 50 | 50 | 10 | 5 | 5 | 10 | 5 | 255 |
| Load pH | 7 | 5 | 3 | 3 | 10 | 10 | 10 | 10 | 10 | 50 | 50 | 10 | 5 | 5 | 10 | 5 | 203 |
| Elution Buffer volume | 7 | 5 | 3 | 15 | 10 | 10 | 10 | 10 | 10 | 10 | 10 | 10 | 5 | 5 | 10 | 5 | 135 |
| PLW1 pH | 7 | 5 | 3 | 3 | 10 | 10 | 10 | 10 | 10 | 10 | 10 | 10 | 5 | 5 | 10 | 5 | 123 |
| Equilibration buffer pH | 7 | 5 | 3 | 3 | 10 | 10 | 10 | 10 | 10 | 10 | 10 | 10 | 5 | 5 | 10 | 5 | 123 |
| Equilibration buffer conductivity | 7 | 5 | 3 | 3 | 10 | 10 | 10 | 10 | 10 | 10 | 10 | 10 | 5 | 5 | 10 | 5 | 123 |
| Equilibration buffer volume | 7 | 5 | 3 | 3 | 10 | 10 | 10 | 10 | 10 | 10 | 10 | 10 | 5 | 5 | 10 | 5 | 123 |
| Equilibration buffer flow rate | 7 | 5 | 3 | 3 | 10 | 10 | 10 | 10 | 10 | 10 | 10 | 10 | 5 | 5 | 10 | 5 | 123 |
| Load temperature | 7 | 5 | 3 | 3 | 10 | 10 | 10 | 10 | 10 | 10 | 10 | 10 | 5 | 5 | 10 | 5 | 123 |
| Load conductivity | 7 | 5 | 3 | 3 | 10 | 10 | 10 | 10 | 10 | 10 | 10 | 10 | 5 | 5 | 10 | 5 | 123 |
| PLW2 pH | 7 | 5 | 3 | 3 | 10 | 10 | 10 | 10 | 10 | 10 | 10 | 10 | 5 | 5 | 10 | 5 | 123 |
| PLW2 conductivity | 7 | 5 | 3 | 3 | 10 | 10 | 10 | 10 | 10 | 10 | 10 | 10 | 5 | 5 | 10 | 5 | 123 |
| Elution Buffer conductivity | 7 | 5 | 3 | 3 | 10 | 10 | 10 | 10 | 10 | 10 | 10 | 10 | 5 | 5 | 10 | 5 | 123 |
| PEW Buffer pH | 7 | 5 | 3 | 3 | 10 | 10 | 10 | 10 | 10 | 10 | 10 | 10 | 5 | 5 | 10 | 5 | 123 |
| PEW Buffer conductivity | 7 | 5 | 3 | 3 | 10 | 10 | 10 | 10 | 10 | 10 | 10 | 10 | 5 | 5 | 10 | 5 | 123 |
| PEW Buffer Volume | 7 | 5 | 3 | 3 | 10 | 10 | 10 | 10 | 10 | 10 | 10 | 10 | 5 | 5 | 10 | 5 | 123 |
| PEW Buffer flow rate | 7 | 5 | 3 | 3 | 10 | 10 | 10 | 10 | 10 | 10 | 10 | 10 | 5 | 5 | 10 | 5 | 123 |

**Detailed Integrated Lonza templates & knowledge available for all unit operations at varying sites and scales i.e. UK, USA, Singapore & Spain.**

PharmaBiotech

# QbD across Development Lifecycle
# Risk Evaluation Stage – Use of Design of Experiments (DOE)

Lonza

## SOUND SCIENCE - Risk Evaluation: Characterisation/Design Space Construction

Lonza

**Central Composite Design (CCD)**

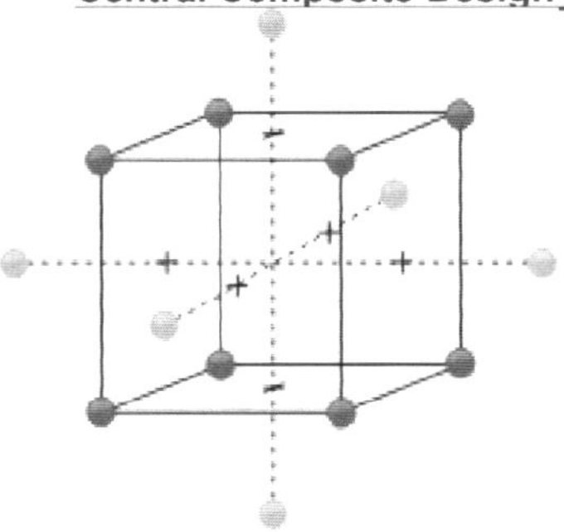


| * | No. Experiments | | Efficiency | |
|---|---|---|---|---|
| Variables (k) | CCD | DD | CCD | DD |
| 3 | 15 | 13 | 0.67 | 0.77 |
| 4 | 25 | 21 | 0.6 | 0.71 |
| **5** | **43** | **31** | **0.49** | **0.68** |
| 6 | 77 | 43 | 0.36 | 0.65 |
| 7 | 143 | 57 | 0.25 | 0.63 |
| 8 | 273 | 73 | 0.16 | 0.62 |

**Doehelrt Design (DD)**

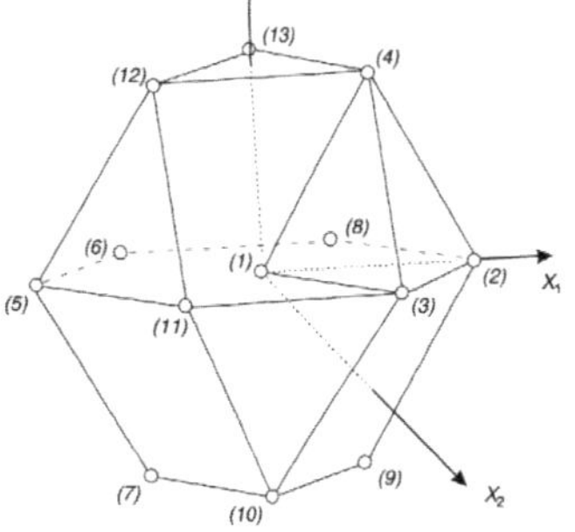


- Classic cube construction for 3 factors.
- Commonly applied.
- Simpler Geometry.
- $2^k + 2k$ + #(centre points) = no. runs

- Cuboctahedron for 3 factors.
- Less commonly applied.
- Hex. Shape can be augmented.
- $K^2 + k$ + #(centre points) = no. runs

## SOUND SCIENCE
## Risk Evaluation – Benefit of DOE approach

Lonza

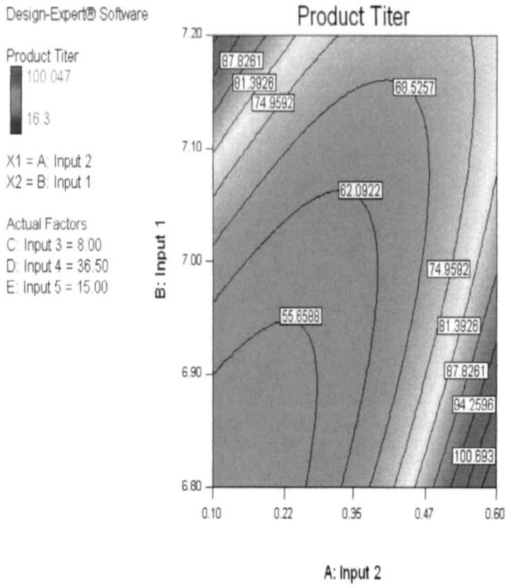


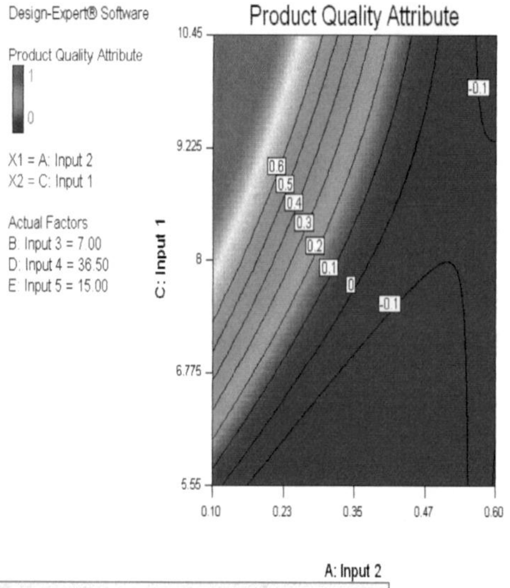


Characterisation space allows for identification of multiple interaction effects (synergistic & antagonistic) & selection of appropriate operating design space.

## Design Space Implementation

Lonza

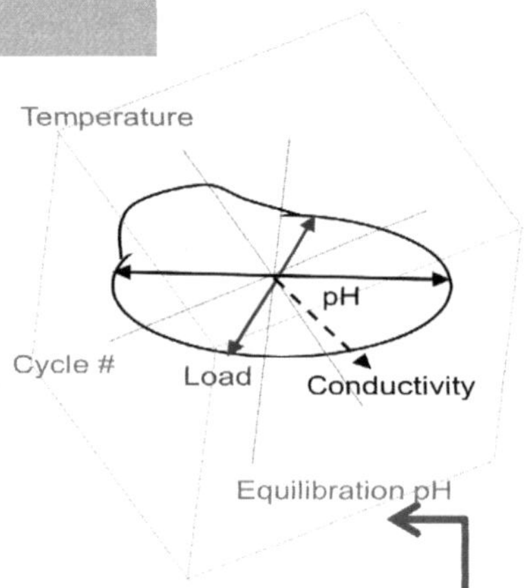


Combination of Univariate and Multivariate (DOE) small and large scale data used to determine appropriate parameter settings for each input.

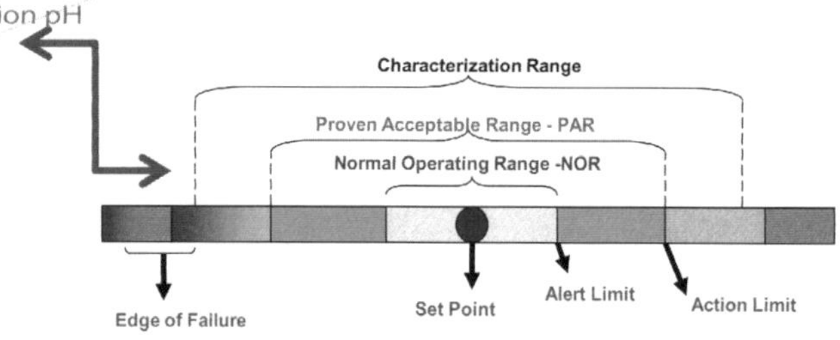

## Cell Count Design Space - Seeding Vessel (N-1)

LONZA

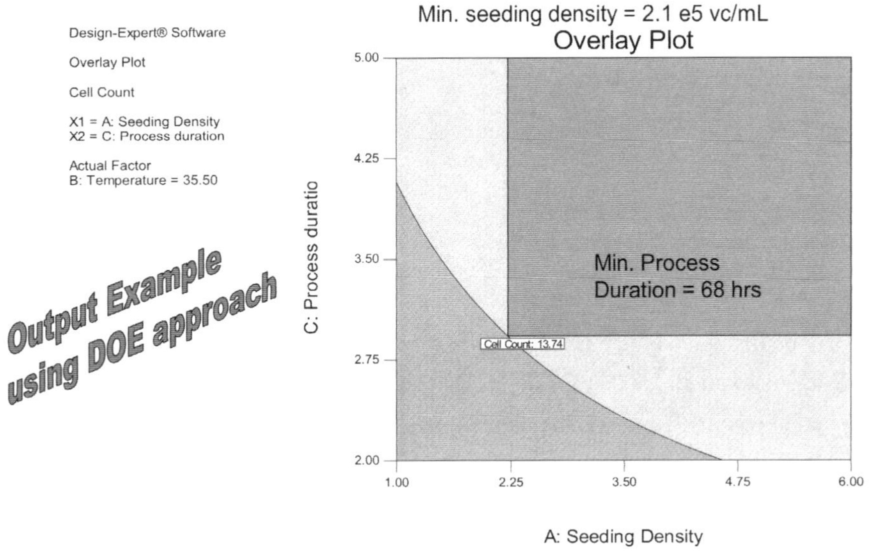


LONZA

## Scientific based quality risk management:
### Phase 1: Data Gathering

**1. Risk Identification** → **2. Risk Analysis** → **3. Risk Evaluation**

1. Risk Identification:
- ✓ Prior knowledge
- ✓ Process maps
- ✓ Cause & Effect matrix

2. Risk Analysis:
- ✓ FMEA – equipment, cause & effect
- ✓ Finalisation of exp. plan

3. Risk Evaluation:
- ✓ DOE/OFAT expt.
- ✓ Design space determination

Use of QRM toolbox

Appropriate SME input from all impacted parties critical.

Experimentation

As per ICH Q9

PharmaBiotech

## QbD across Development Lifecycle
## Phase 2 – Data Interpretation & Demonstration
## (a) Parameter Designation (b) Process Validation

LONZA

Only Applicable for Products in Final Stages of Clinical Development

## Scientific based quality risk management
## Phase 2: Data Interpretation & demonstration

LONZA

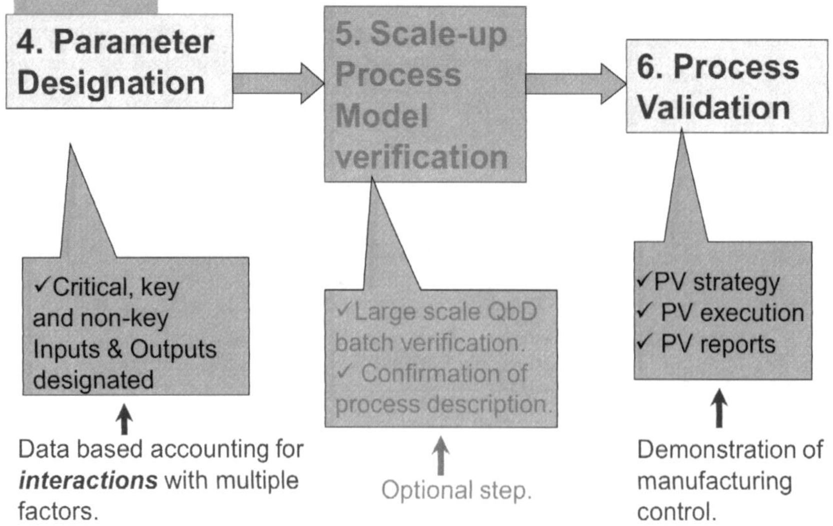

## Step 4. Parameter Designation

Lonza

- CQA – Critical quality attribute
- QA – Quality attribute
- CPP – Critical process parameter
- KPP – Key process parameter
- Non-KPP – Non Key process parameter

(CQA, QA: Outputs; CPP, KPP, Non-KPP: Inputs)

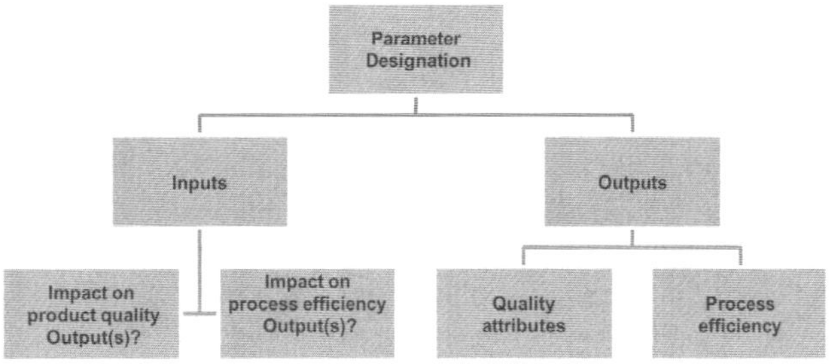


## Step 4. Parameter Designation – Decision tree for Inputs

Lonza

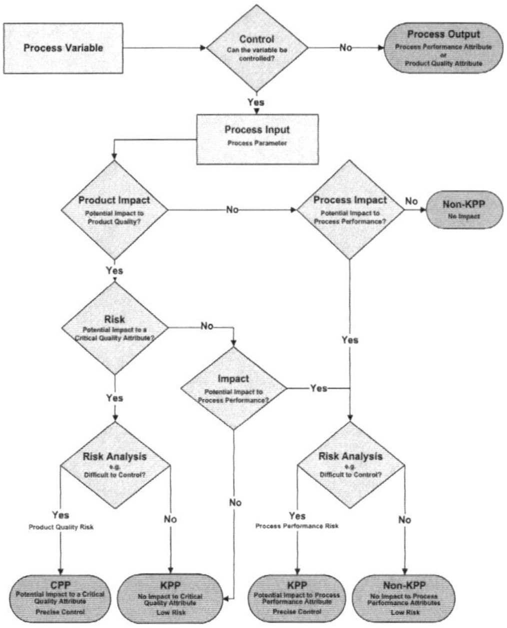

## Step 4: Parameter Designation

Lonza

### Production bioreactor step example

- Cause & Effect matrix

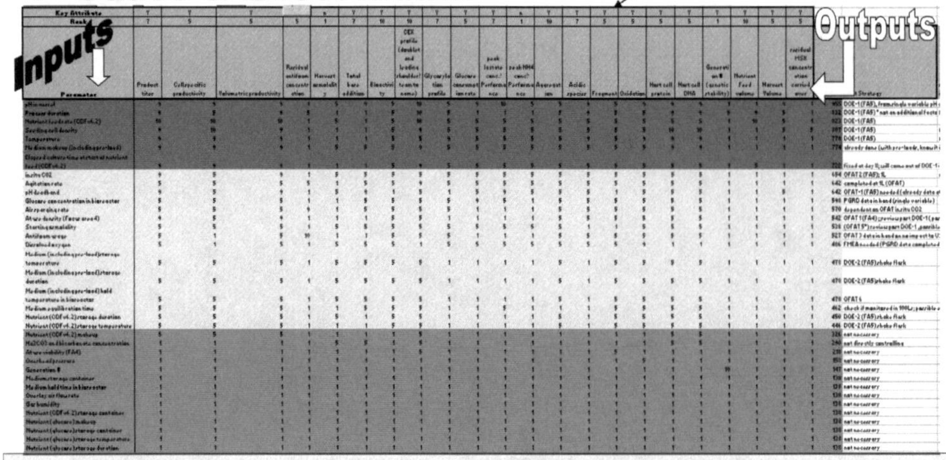


Risk Identification: "Systemic use of information referring to the risk question or problem description."
As per ICH Q9

Lonza

## Quality Risk Management Example - FMEA

### e.g. Vial thaw & Initial shake flask stage:

| | Parameter Name | Failure mode | Effects of failure mode | Severity score | Occurrence | Detectability | RPN | Rationale |
|---|---|---|---|---|---|---|---|---|
| 1 | | | | | | | | |
| 16 | **Vial Thaw - heater block** | | | | | | | |
| 17 | Thaw temperature | High | Cell damage | 5 | 3 | 3 | 45 | LPQ data supports designated NOR, WFI is pre-warmed in a water bath at controlled temperature set point |
| 18 | | Low | Extended thaw time | 5 | 3 | 3 | 45 | |
| 19 | Vial thaw time | High | Cell damage | 9 | 3 | 1 | 27 | No vial thaw failures observed to date and considered robust cell line. LPQ data supports vial thaw time, and combined thaw and transfer time as specified in BR. |
| 20 | | Low | Incomplete thaw - | 9 | 3 | 1 | 27 | |
| 21 | **Primary Shake Flask (125mL)** | | | | | | | |
| 22 | Incubator Temperature | High | viability and final VCD | 9 | 3 | 1 | 27 | Incubators tightly controlled and alarmed and LPQ data to support designated ranges |
| 23 | | Low | | 9 | 3 | 1 | 27 | |
| 24 | Primary shake flask (125mL) working volume | High | inadequate mass transfer | 5 | 3 | 7 | 105 | Standard site volumes for 125mL shake flask |
| 25 | | Low | increased foaming | 5 | 3 | 7 | 105 | |
| 26 | Incubator $CO_2$ conc | High | impact media pH and culture performance | 5 | 3 | 3 | 45 | Standard gassing procedure for Lonza, historical experience to support numerous cell lines |
| 27 | | Low | | 5 | 3 | 3 | 45 | |
| 28 | Target seed density - not counted after seeding | High | viability and final VCD | 9 | 5 | 7 | 315 | Parameter is target seed density so not verified with post-inoc count. Selected as non KPP as at use cell density and final viability designated as KPA's which must be met prior to forward processing |
| 29 | | Low | | 9 | 3 | 7 | 189 | |
| 30 | Culture Duration | High | viability and final VCD | 9 | 1 | 1 | 9 | Culture duration controlled with LPQ data to support wide culture duration |
| 31 | | Low | | 9 | 1 | 1 | 9 | |
| 32 | Agitation Rate of rocker base | High | viability and final VCD | 5 | 3 | 3 | 45 | Rocker rate controlled and monitored |
| 33 | | Low | | 5 | 5 | 3 | 75 | |
| 34 | Agitation platform orbital throw | High | viability and final VCD | 5 | 1 | 5 | 25 | Site standard equipment |
| 35 | | Low | | 5 | 1 | 5 | 25 | |
| 36 | Final Cell Viability (prior to subculture) | | | | | | | |

Risk Analysis: "is the estimation of risk with the identified hazard."
As per ICH Q9

## Case Study: Parameter Designation Inputs

Lonza

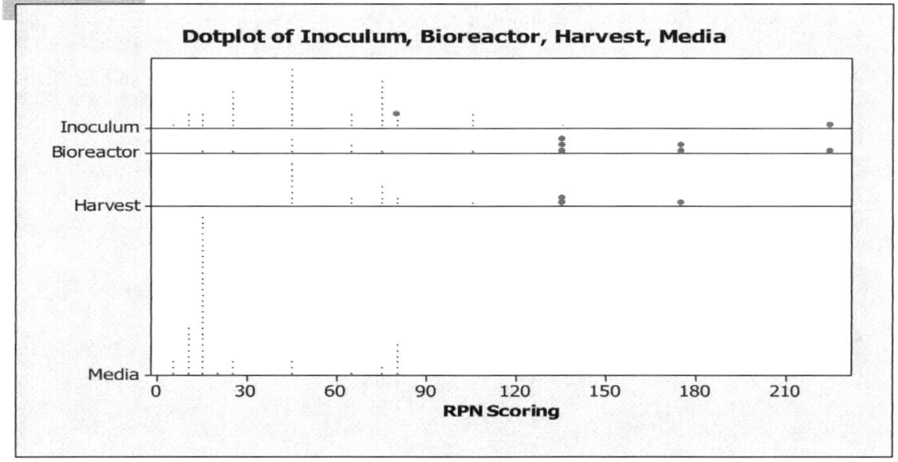


For the Upstream and Downstream stages *(data not shown)*, the higher RPN scores (≥81) were typically designated as being KPPs or CPPs as indicated by the red symbols. The majority of parameters scored < 81 indicating appropriate manufacturing ranges and control with low risk to product quality or process performance. These were typically designated as being non-key.

## Step 6. Process Validation

Lonza

- Define Process validation strategy & successfully execute.

Established:

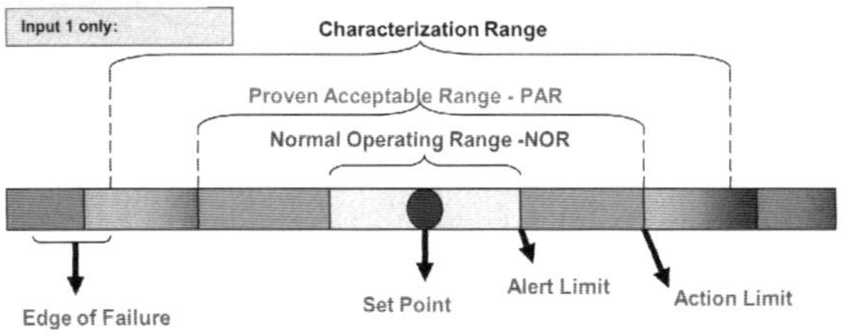


"Conformance batches are prepared to demonstrate that, under normal operating conditions and defined ranges of operating parameters, the process appears to make acceptable product. Prior to the manufacture of the conformance batches the manufacturer should have identified and controlled all critical sources of variability." **FDA policy document 490.100**

Design Space:

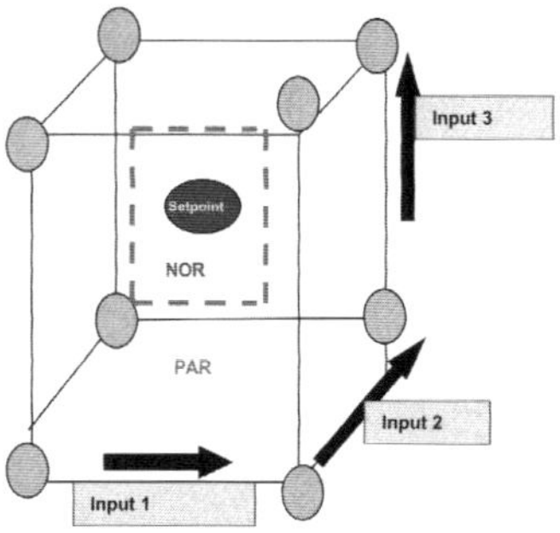

## Scientific based quality risk management
## Phase 2: Data Interpretation & demonstration

Lonza

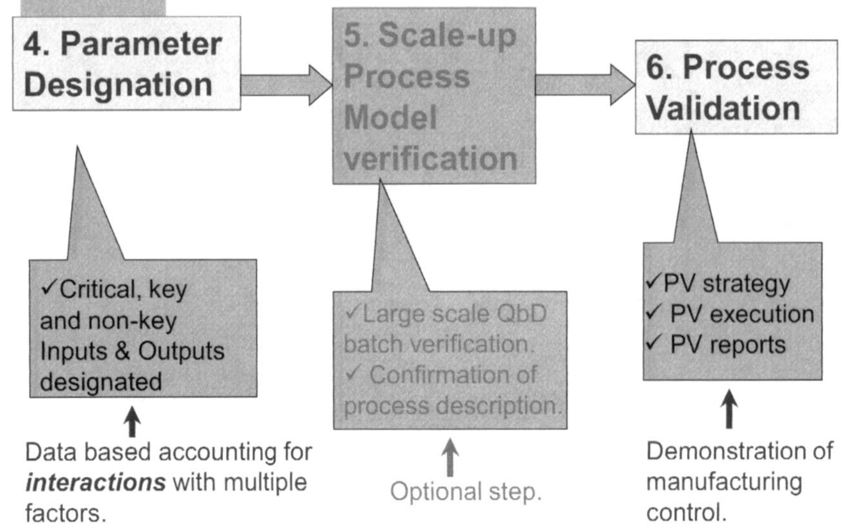


PharmaBiotech

## QbD across Development Lifecycle
## Phase 3 – BLA submission

Lonza

## Sound Science & Scientific based Quality risk management – Ultimate Aim: BLA submission

LONZA

***__Documents from early stages are iterative__*** such that the initial templates are developed and refined with improved process knowledge & greater understanding of the manufacturing procedures & capabilities.

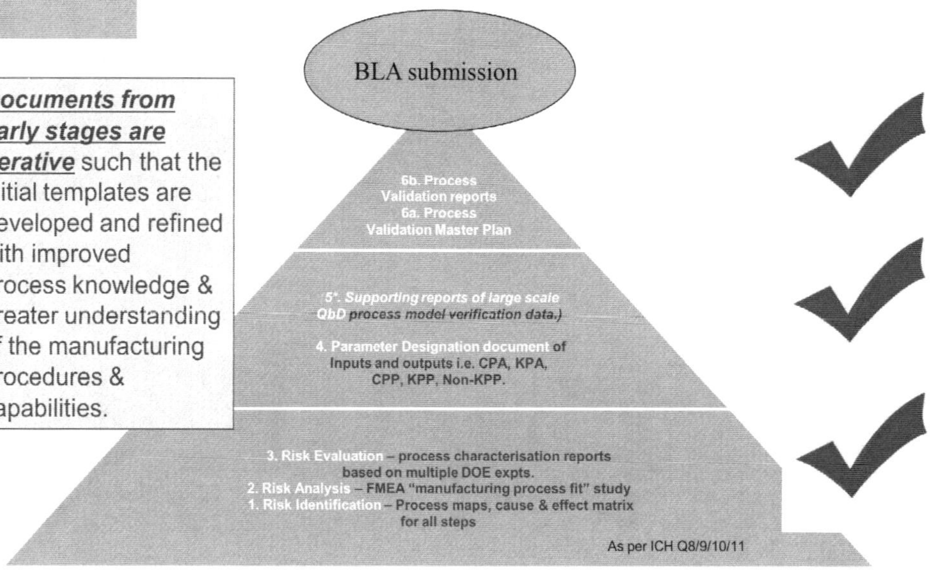


* Not a requirement/specified in guidance documents.

PharmaBiotech

## CMO Perspective – Lessons Learned

LONZA

## Development & Manufacture of Drug Substance in the Development Lifecycle

**Scope & objectives are dependent on the stage of development:**

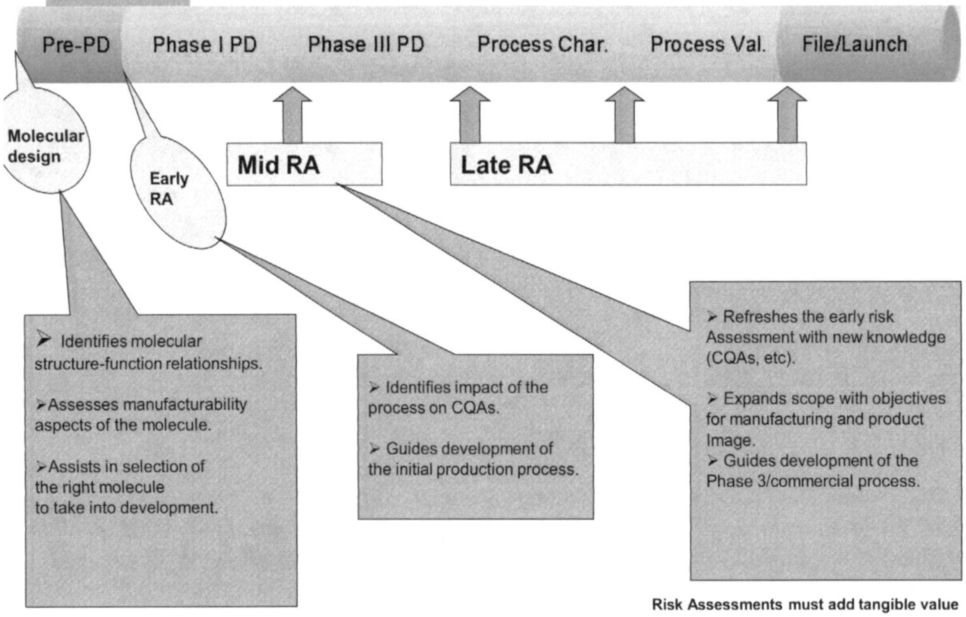


Risk Assessments must add tangible value

## Elements of "Islands of Automation" - Process Development Workflow

Lonza

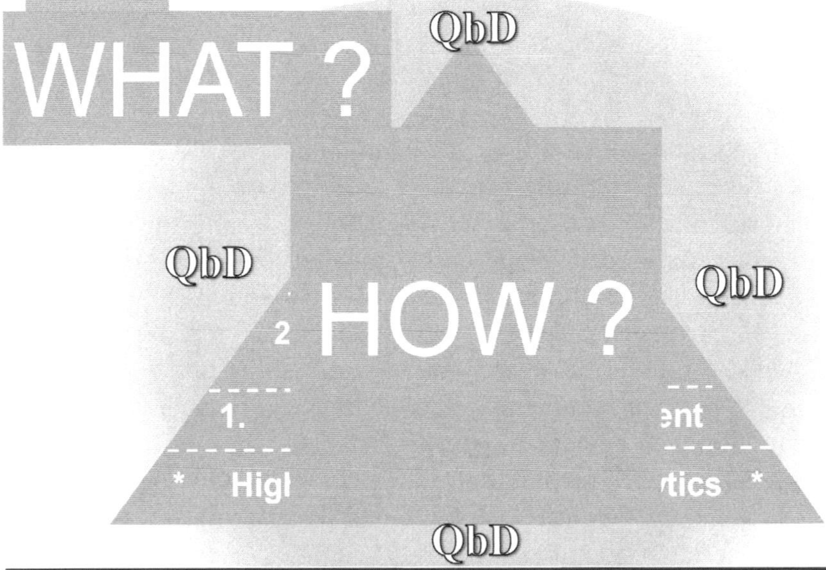


Development of Cell Culture and Purification Process as Single Entity

Lonza

## Acknowledgements

- Cell Culture Development
- Process Sciences
- Analytical Services
- Purification Development
- Manufacturing Sciences And Technology
- Operations

46
Mrz-13

**Using a Statistical and Risk Based Approach to Improve Process Robustness and Control in a Late Stage Antibody Process**

Conley, L., Biogen IDEC
Originally presented on: 2/26/2013 11:35 - 12:05

This presentation describes an approach for improving robustness and enhancing control of a late stage antibody process. Traditional approaches have usually focused on developing independent statistical models for each individual process step. This limits the understanding of overall process robustness and the inter-relationship between process steps. A more holistic approach was used in the development of this late stage antibody process. A risk based approach identified process parameters that may influence quality attributes, which were then used to construct statistical experimental designs. These designs were used to characterize and create predictive models for each step of the process from cell culture through final ultrafiltration/diafiltration. Most of the impurities generated in cell culture were independently removed by the various downstream purification steps with minimal impact from the load levels in the clarified cell culture fluid or eluates from a previous chromatography step. However, a single type of impurity was identified as the having the greatest potential impact on the robustness of the manufacturing process because the impurity levels varied considerably during cell culture. Process parameters affecting this impurity generation and removal were holistically evaluated through the entire process, critical parameters identified, and effective controls established to regulate the impurity levels. This overall approach resulted in a more complete understanding of the design space around each process step and the capability of the process as a whole to clear the impurity to an acceptable level.

**Quality by Design – QbD at a CMO**

Abhinav A. Shukla, Ph.D.
Vice President, Process Development & Manufacturing
KBI Biopharma, Durham NC

# Quality by Design (QbD)

- "Quality by design means designing and developing manufacturing processes during the product development stage to consistently ensure a predefined quality at the end of the manufacturing process." ICH Q10, FDA 2006

Process Design (Process Development) → Process Control Strategy Definition → Process Validation → Continued Process Verification → Process Design (Process Development)

KBI BIOPHARMA

-Confidential-

THE SCIENCE TO SUCCEED.

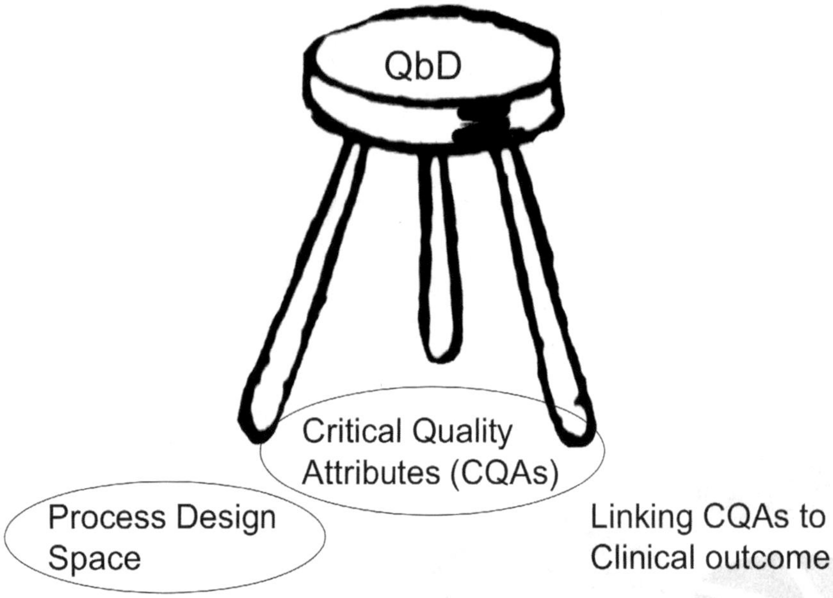


THE SCIENCE TO SUCCEED.

## Development Phase

- Utilizing the right set of analytical tools for in-process testing and release
- Characterization assays are equally important
- Utilizing a broad set of tools up front gives the best chance of determining CQAs & linking them to the process

THE SCIENCE TO SUCCEED.

## Comprehensive Analytics

### Analytical Methods Portfolio

- **Protein Primary Structure**
  - Peptide Sequencing via LC/MS/MS
  - Amino Acid Analysis
  - Peptide Mapping
- **Biophysical Characterization**
  - CD, FTIR, DSC, DLS, fluorescence spectroscopy
- **Capillary and Slab Gel Electrophoresis**
  - CZE
  - SDS-CGE
  - cIEF and icIEF
  - SDS-PAGE and IEF
  - Western blot
  - Microchip electrophoresis
  - 2D gels and blots
- **Glycan Analysis**
  - Oligosaccharide mapping
  - Monosaccharide composition
  - Sialic Acid Quantitation
- **Process Residuals**
  - ELISA (HCP, protein A etc.)
  - HPLC (antibiotics, IPTG, detergents, etc)
  - qPCR (DNA)
- **HPLC**
  - Size Exclusion (with MALLS)
  - Ion Exchange
  - Reverse Phase
  - Hydrophobic Interaction
  - Affinity
- **Potency Assays**
  - Binding Assays via ELISA, Biacore and ForteBio
  - Cell Based Assays (e.g., proliferation, cytokine release, etc.)
- **Mass Spectrometry**
  - Intact mass
  - Peptide mapping with LC/MS or LC/MS/MS
  - Disulfide Mapping
  - Post translational modifications (e.g., oxidation, deamidation)
  - PEGylation site identification
  - Glycan Identification & site identification
- **Particle measurements**
  - Visible & sub-visible particles

KBI BIOPHARMA -Confidential- THE SCIENCE TO SUCCEED.

## Development phase goals

- Developing a robust and reproducible process with high productivity
  - Single development cycle
  - Collecting data linking CQAs to process
- Process documentation
  - IND enabling process documentation
  - Mapping influence of process parameters on outcome

KBI BIOPHARMA THE SCIENCE TO SUCCEED.

# Designing more efficient HCP clearance into the downstream process

- Most current chromatographic steps are designed to remove impurities based on differential binding to the stationary phase surface
- Conventional wisdom: wash conditions are between binding and elution conditions
- Orthogonal approach → disrupt impurity-product interactions

KBI BIOPHARMA

THE SCIENCE TO SUCCEED.

# Enhancing HCP clearance across Protein A

- HCPs form a diverse set of impurities
- HCP clearance is a key concern in biopharmaceutical separation processes

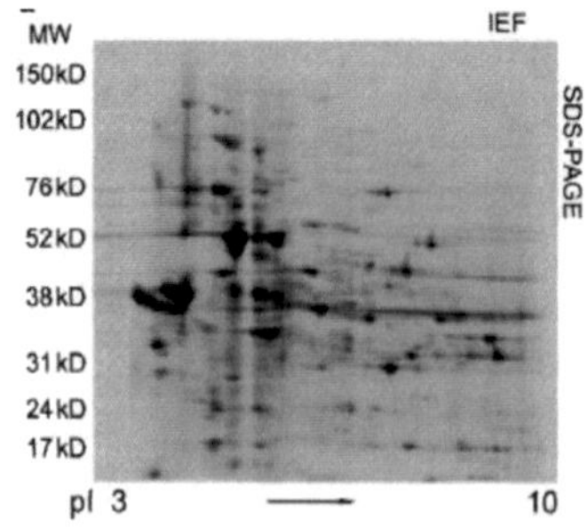


KBI BIOPHARMA

THE SCIENCE TO SUCCEED.

8

## Enhancing HCP clearance across Protein A

Do HCPs co-elute with the product or co-associate with the product?

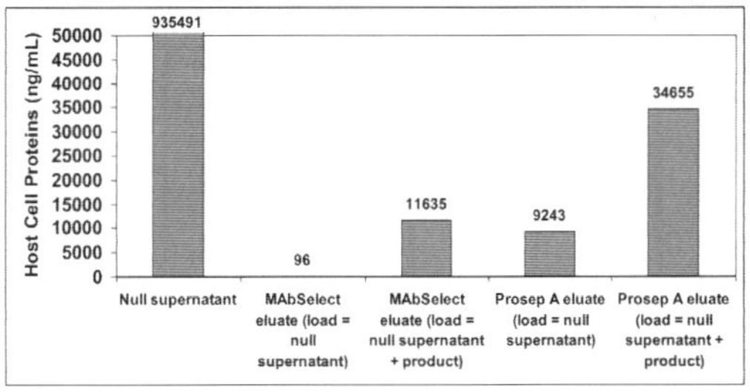


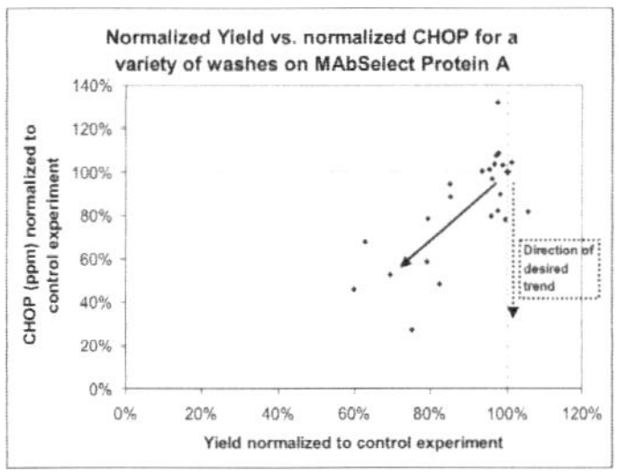


Washes can be developed to disengage HCPs from the product rather than disrupt product-Protein A ligand interactions

Biotechnology Progress, 24, 1115-1121, 2008.

KBI BIOPHARMA -Confidential- THE SCIENCE TO SUCCEED.

## Enhancing HCP clearance across Protein A

- Use washes at high pH (pH > 7) to preserve Protein A – mAb interactions
- Include selective modulators (moderate concentrations of urea, ethylene glycol, salts, arginine) in washes to disrupt HCP-mAb interactions

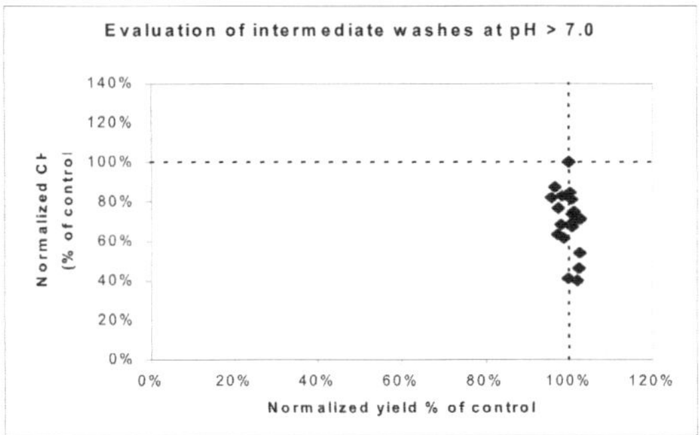


KBI BIOPHARMA THE SCIENCE TO SUCCEED.

# Mixed Mode Chromatography

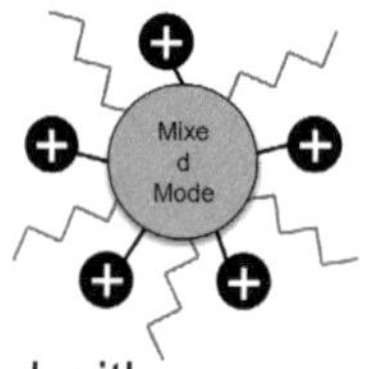


- Takes advantage of more than one type of interaction
- Can reduce process steps
- Provides enhanced selectivity, "pseudo-affinity"
- Several mixed mode resins have recently been developed with:
  - » *Increased loading capacities*
  - » *Higher ionic strength tolerance*

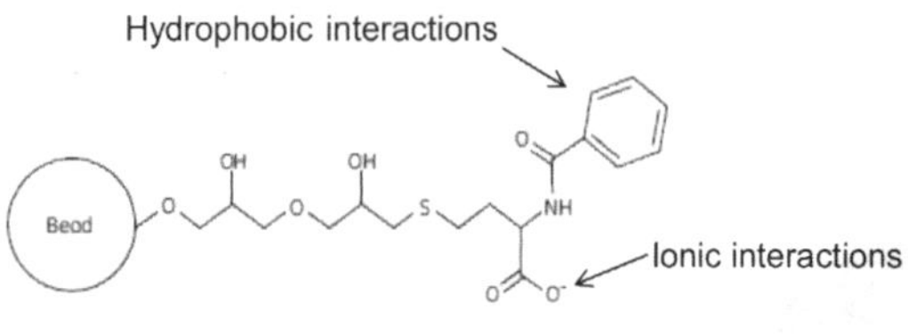


**GE Healthcare, Capto MMC ligand**

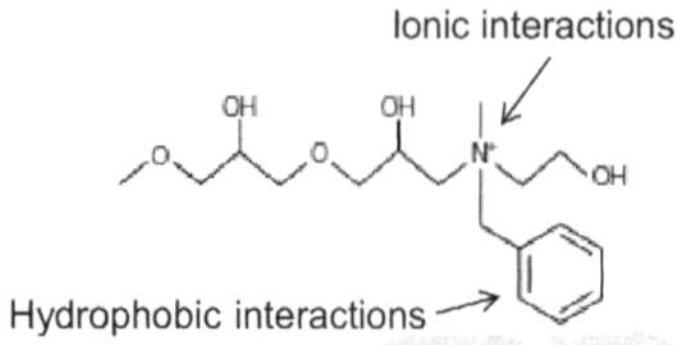


**GE Healthcare, Capto Adhere ligand**

KBI BIOPHARMA

THE SCIENCE TO SUCCEED.

# Log k' vs Log [NaCl]

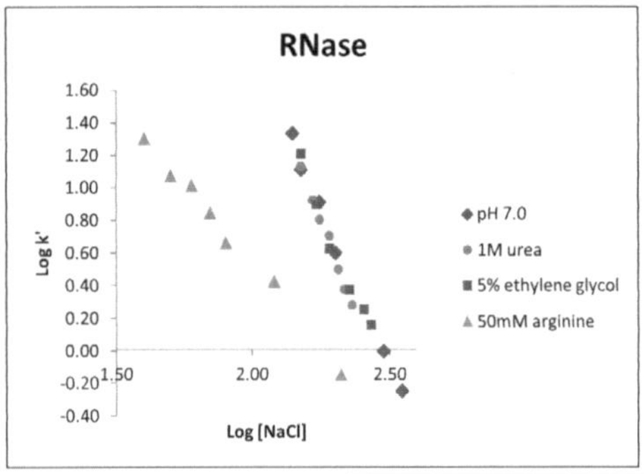


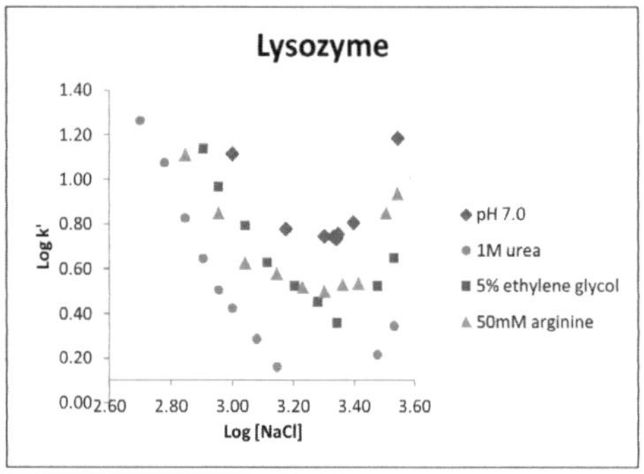


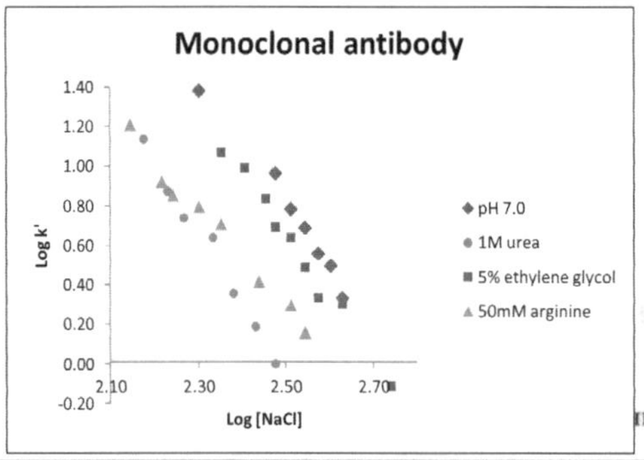


KBI BIOPHARMA

THE SCIENCE TO SUCCEED.

# Wash development on mixed mode

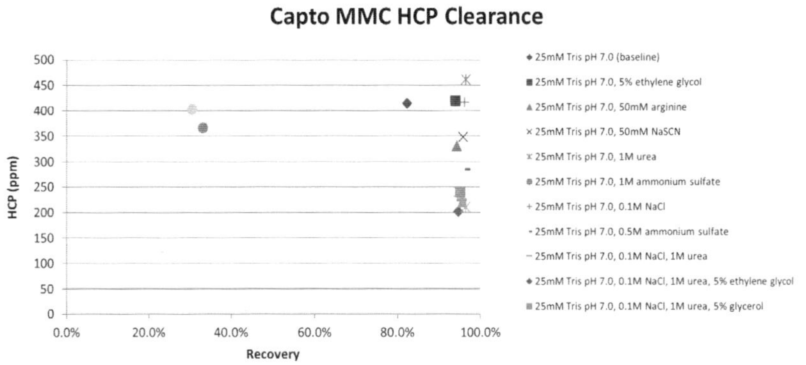


- Selective wash strategies can eliminate one chromatographic step in non-mAb processes
- Designing quality into the process

KBI BIOPHARMA

THE SCIENCE TO SUCCEED.

# Process characterization phase

- Design and execution of scale-down studies to study the impact of process parameters on CQAs
  - Define process control strategy
  - Identify CPPs and develop information set on other process parameters

Characterization range ≥ Operating range

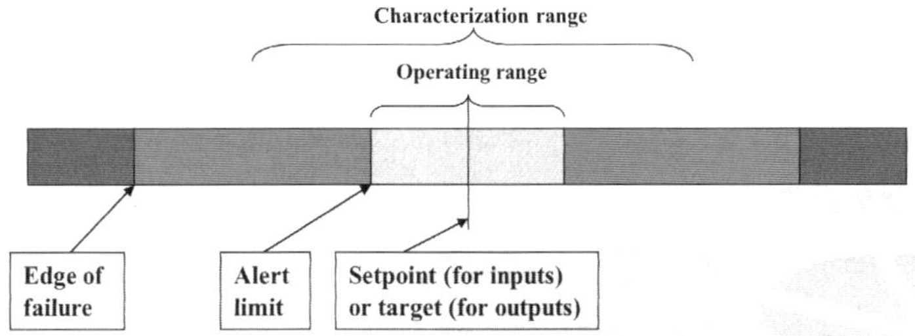


KBI BIOPHARMA

THE SCIENCE TO SUCCEED.

# Process Design Space

Design Space (ICH Q8, 2006): The multidimensional combination and interaction of input variables (e.g., material attributes) and process parameters that have been demonstrated to provide assurance of quality.

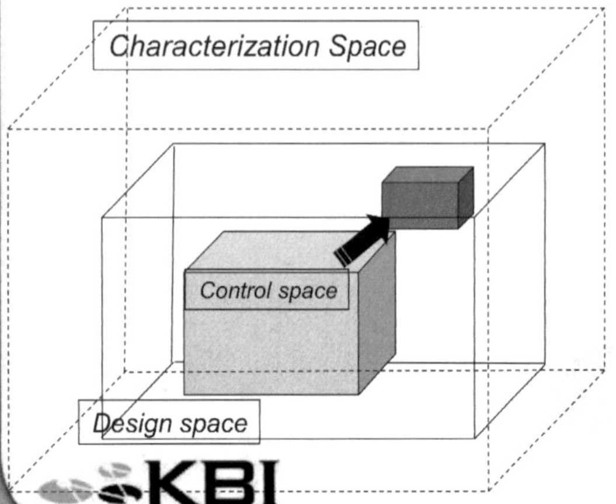


- Higher level of assurance of product quality
- Manufacturing Efficiency and Flexibility
  - Continuous process improvement while maintaining product quality

KBI BIOPHARMA

-Confidential-

THE SCIENCE TO SUCCEED.

# Classification of operating parameters

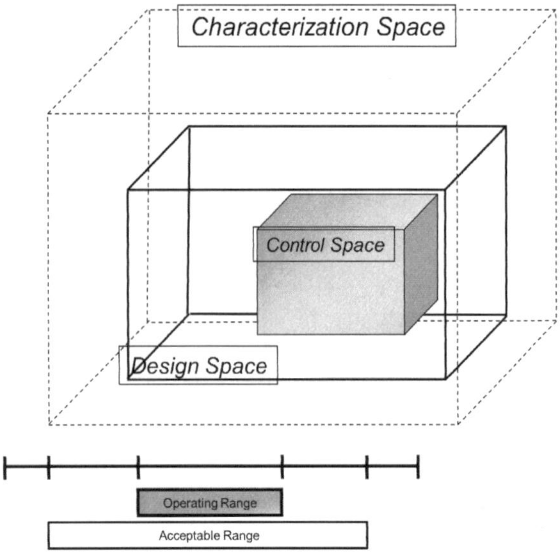


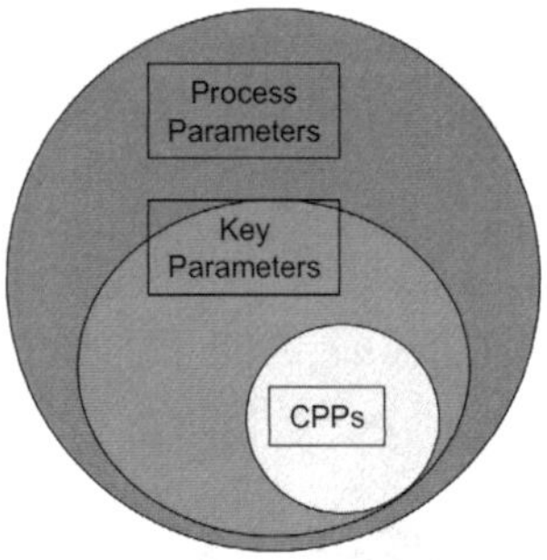



KBI BIOPHARMA

-Confidential-

THE SCIENCE TO SUCCEED.

## Integrative Approach

### Each step is influenced by the preceding step

- Shake flask and seed bioreactor parameters may affect growth rate in the seed bioreactor.
- Seed bioreactor and production bioreactor parameters may affect productivity and critical quality attributes.
- Production bioreactor parameters may affect downstream steps.
- Characterization studies are linked.

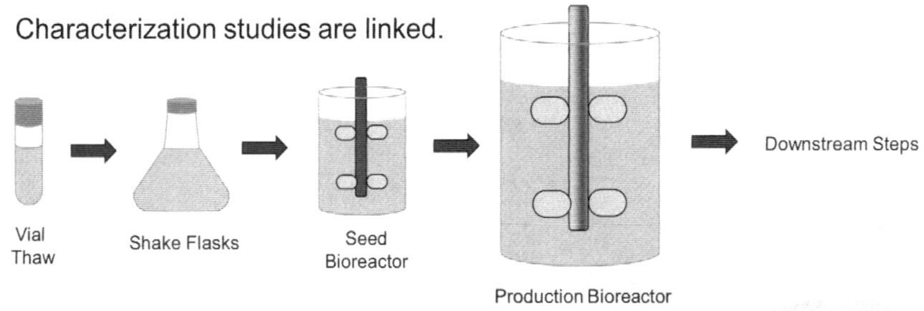


Biotechnology and Bioengineering, 106(6), 894-905, 2010.

KBI BIOPHARMA -Confidential- THE SCIENCE TO SUCCEED.

## Production Bioreactor

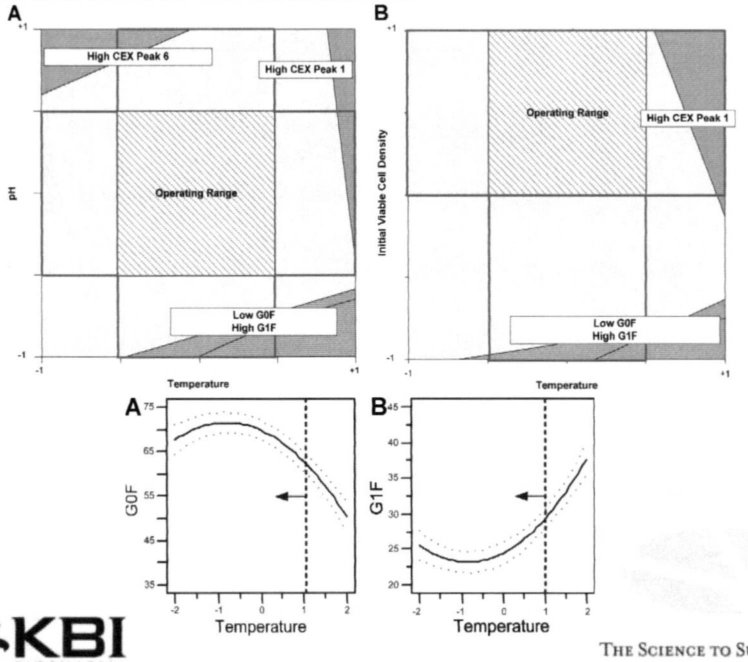


KBI BIOPHARMA THE SCIENCE TO SUCCEED.

# Establishing A Process Control Strategy

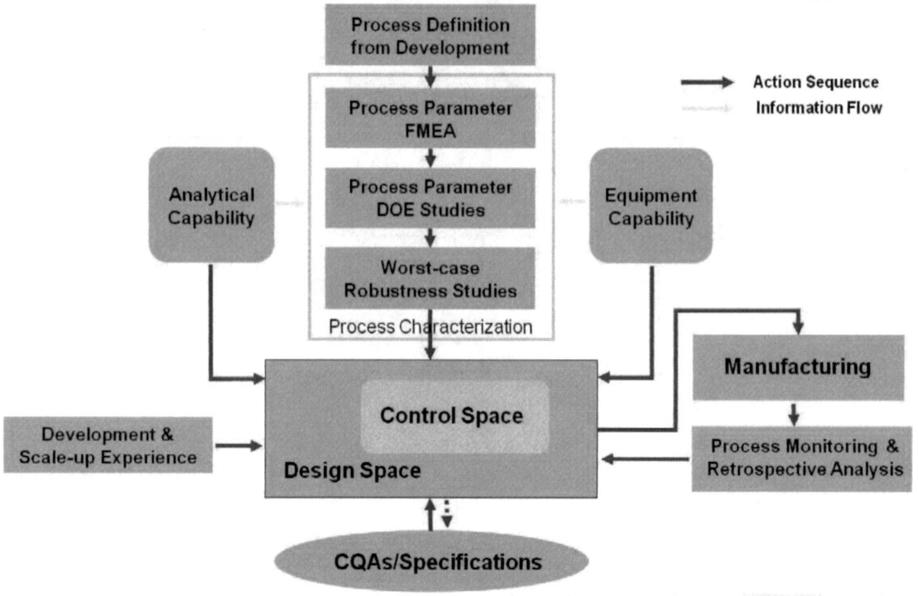


-Confidential-

The Science to Succeed.

# Application of Quality by Design (QbD): downstream resin lot variability

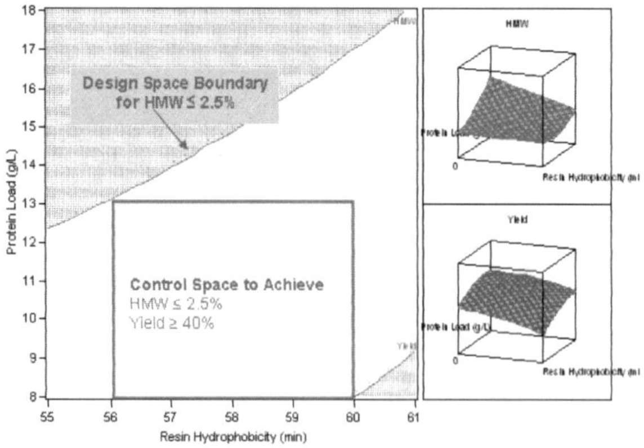


Biotechnology and Bioengineering, 107(6), 989-1001, 2010.

-Confidential-

The Science to Succeed.

## Evolving expectations in Process Validation

- Q7A definition: "Process validation is the documented evidence that the process, operated within established parameters can perform effectively and reproducibly to produce an intermediate or API meeting its predetermined specifications and quality attributes"
- FDA guidance, Jan 2011: "The collection and evaluation of data, from the process design stage through commercial production, which establishes scientific evidence that a process is capable of consistently delivering quality product"
- Process validation is now viewed as a process that occurs throughout the lifecycle of a product

Process Design (Process Development) → Process Control Strategy Definition → Process Qualification → Continued Process Verification → Process Design (Process Development)

KBI BIOPHARMA

THE SCIENCE TO SUCCEED.

-Confidential-

## Scale-Down Process Validation Studies

- Scale-down validation studies in addition to large-scale process validation (conformance lots)
- Probe extremes in the process and demonstrate them to be acceptable
- Examples
  - Reprocessing validation – combine hold times with process conditions that create the greatest stress on the protein
  - Intermediate hold times – combine hold times and demonstrate releasable drug substance
  - Viral clearance studies
  - Impurity clearance studies

KBI BIOPHARMA

THE SCIENCE TO SUCCEED.

# Validation of Host Cell Protein Clearance

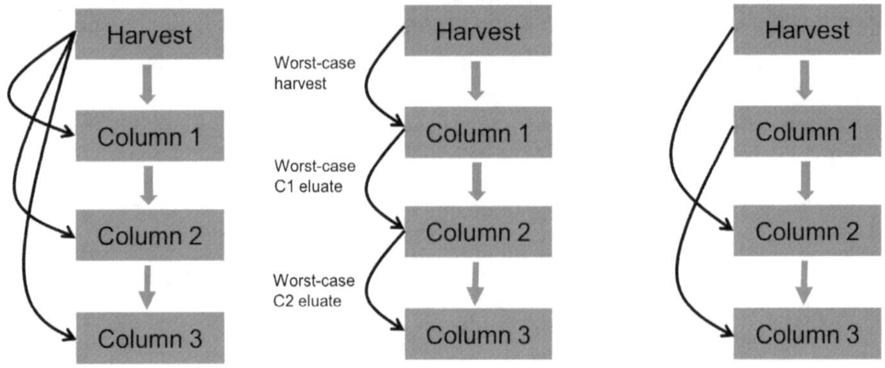


### Spiking Strategy

- Some CHOP species in harvest may not be encountered by C2 and C3 in Mfg
- LVR could be overstated for C2 and C3

### Worst-case Strategy

- CHOP species in eluate is relevant to the next step
- More accurate evaluation of LRV
- Need process characterization to identify worst-case condition

### By-pass Strategy

- HCP species in load are relevant to that process step in case the previous step is by-passed (e.g. "resin bed channeling")
- Represents most "challenged" scenario

*Biotechnol. Progr.*, 24(3), 615 – 622, 2008

KBI BIOPHARMA

-Confidential-

THE SCIENCE TO SUCCEED.

# Process Analytical Technology

- PAT: "a system for designing, analyzing and controlling manufacturing through timely measurements (i.e. during processing) of critical quality and performance attributes of raw and in-process materials and processes, with the goal of ensuring final product quality" FDA Guidance

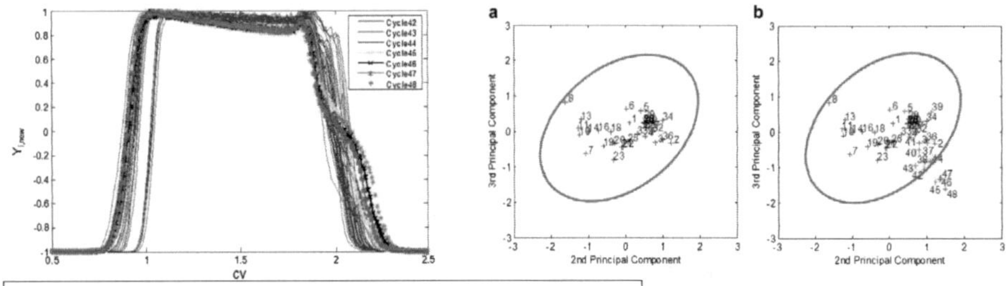


Biotechnology and Bioengineering, 108(1), 59-68, 2011.

KBI BIOPHARMA

THE SCIENCE TO SUCCEED.

# Biologics Commercialization

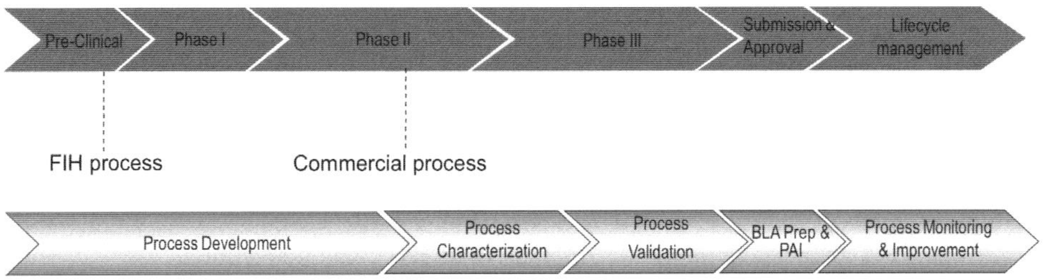


FIH Process
- Deliver clinical process quickly
- Platform process
- Clinical Supply

Commercial Process
- Deliver manufacturing process for registrational trials and market
- Design keeping large-scale manufacturing in mind
- Improve productivity, efficiency, robustness, manufacturability, COGs
- Analytical characterization and method development

Process Characterization and Validation
- Develop IPC strategy through understanding of process inputs and outputs (design space)
- Scale-down characterization and validation studies
- Large-scale process validation to demonstrate process consistency
- BLA preparation
- Supporting documents for licensure inspections
- Post-commercial process improvements (CI)
- Post-commercial process monitoring

Gottschalk U., Konstantinov K., Shukla A. Nature Biotechnology, 30(6), 489-491, 2012

KBI BIOPHARMA -Confidential- THE SCIENCE TO SUCCEED.

# Why a CMO can lead to a stronger QbD package

- Expertise in developing, scaling up and operating processes for a wide variety of product types (mAbs, enzymes, cytokines, vaccines, novel scaffolds)
- Focused expertise in analytical and process development & cGMP manufacturing
- Can lend experience from other programs so the outcome is not a one-off exercise
- Can complement client's internal efforts on linking CQAs to clinical outcome
- Flexibility towards owning studies vs. executing them
- Can help overcome resource crunch for a client

KBI BIOPHARMA THE SCIENCE TO SUCCEED.

# Conclusions

- QbD guidance has evolved FDA expectations on what is included in licensure filings
- Increased emphasis on process understanding as a basis for control
- Increased emphasis on demonstrating control through scale-down experimentation rather than conformance lots alone
- Increased emphasis on understanding and mapping process design space

KBI BIOPHARMA

THE SCIENCE TO SUCCEED.

GlaxoSmithKline

## *Accelerating Early Phase Development through Efficient Implementation of Molecular Design & Platform Knowledge*

*SBE's 3rd International Conference on Accelerating Biopharmaceutical Development*

*Rick Caimi*



## Overview

- Illustrate embedded Molecule & Process de-risking approach
  - Appropriate medicinal properties/ CMC targets designed into candidates.
- CMC engagement overlaps sig. w/ Discovery prior to CS.
  - Aligns Discovery & Development objectives from "Day 1".
  - Optimum molecule rather than just the one with the fewest "warts"
  - Prevents bad molecules from building significant momentum.
- Acceleration of timelines across portfolio achieved via:
  - Leveraging of knowledge & data across portfolio.
  - Well grounded platform shared across Discovery & Development Org.
  - Prospectively linking methods & processes to long term objectives.
  - Phase appropriate practices
    - Risk based decisions/ resource commitment.
- Application to mAbs and Non-mAbs alike
  - Enable optimal delivery of a portfolio of assets.

GlaxoSmithKline

## BioPharm Strategic Investments

### Embedding MDI/QbD in the Biopharm Development Lifecycle

- "Twinkle to Launch & beyond."
  - Discovery, Development and Manufacturing.
- Modification of both cultural & business practices.
  - An end to the Christmas Candidate!

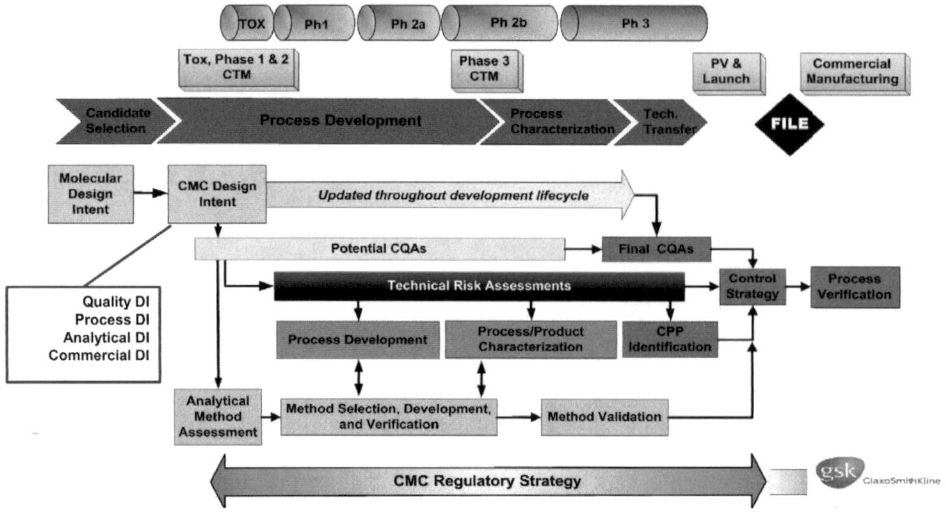


## BioPharm Strategic Investments

### Drivers for MDI/QbD in the Biopharm Development Lifecycle

- Design of optimum molecule.
  - Optimum molecular format/ greater lead diversity explored.
  - Lead satisfy requirements for multiple indications.
  - Qualities enable meaningful differentiation from competitors.
- Ongoing accumulation of **relevant** product knowledge
  - Early identification of CQAs
  - Implementation of suitable control strategies
  - More rapid advancement to commercialization
    - Reduced product attrition
  - Maximize likelihood of clinical/ commercial success.
- Feedback loops – leveraging information across pipeline.

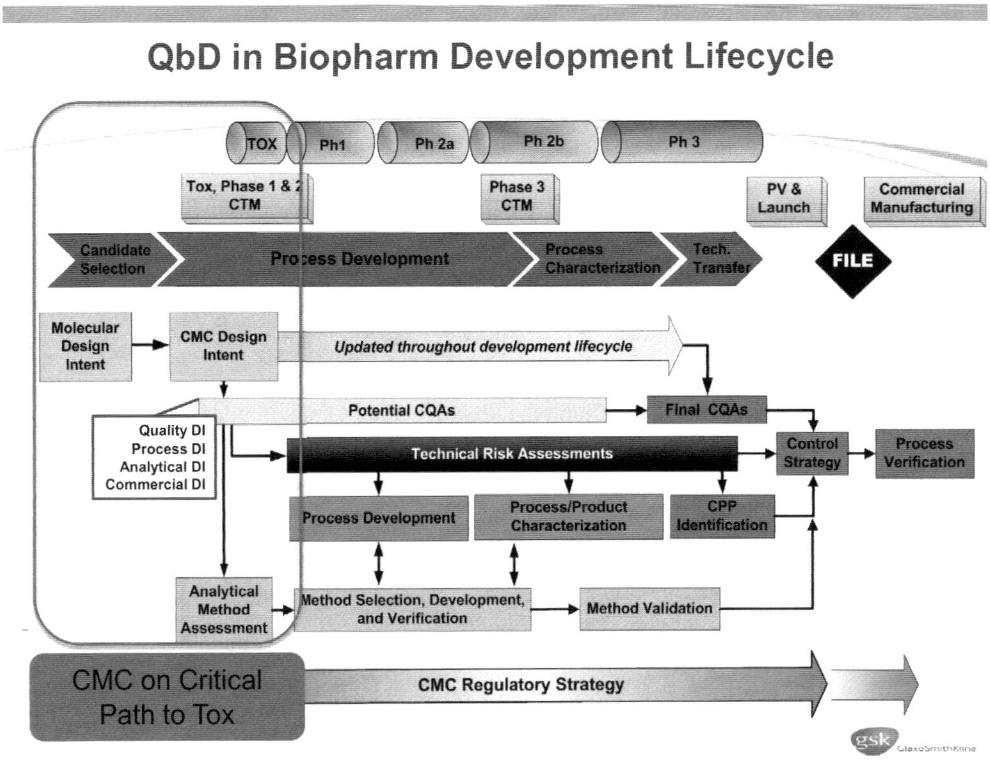
QbD in Biopharm Development Lifecycle
TOX
Ph1
Ph 2a
Ph 2b
Ph 3
Tox, Phase 1 & 2 CTM
Phase 3 CTM
PV & Launch
Commercial Manufacturing
Candidate Selection
Process Development
Process Characterization
Tech. Transfer
FILE
Molecular Design Intent
CMC Design Intent
Updated throughout development lifecycle
Potential CQAs
Final CQAs
Quality DI
Process DI
Analytical DI
Commercial DI
Technical Risk Assessments
Control Strategy
Process Verification
Process Development
Process/Product Characterization
CPP Identification
Analytical Method Assessment
Method Selection, Development, and Verification
Method Validation
CMC on Critical Path to Tox
CMC Regulatory Strategy
gsk GlaxoSmithKline

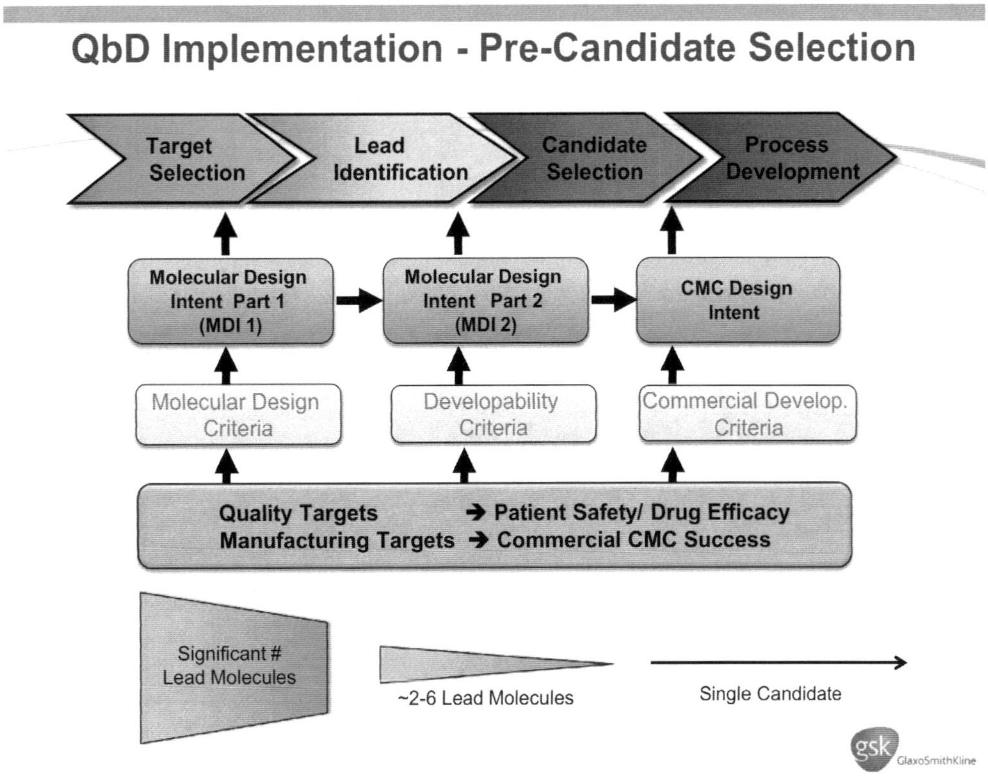
QbD Implementation - Pre-Candidate Selection
Target Selection
Lead Identification
Candidate Selection
Process Development
Molecular Design Intent Part 1 (MDI 1)
Molecular Design Intent Part 2 (MDI 2)
CMC Design Intent
Molecular Design Criteria
Developability Criteria
Commercial Develop. Criteria
Quality Targets → Patient Safety/ Drug Efficacy
Manufacturing Targets → Commercial CMC Success
Significant # Lead Molecules
~2-6 Lead Molecules
Single Candidate
gsk GlaxoSmithKline

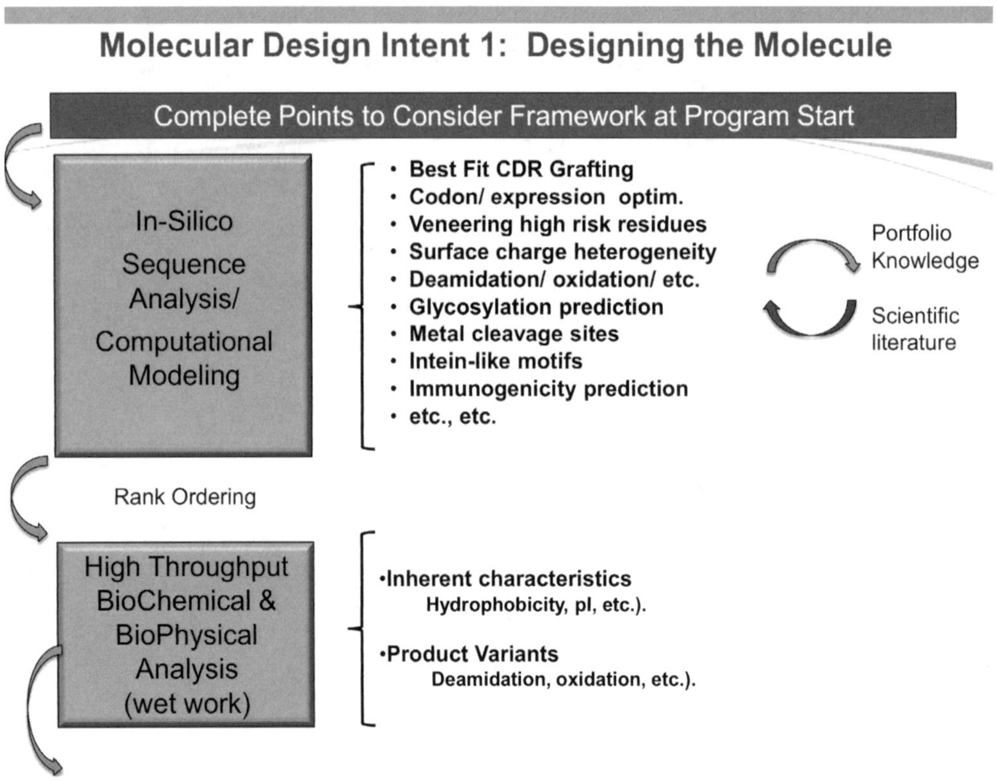
Molecular Design Intent 1: Designing the Molecule
Complete Points to Consider Framework at Program Start
In-Silico Sequence Analysis/ Computational Modeling
• Best Fit CDR Grafting
• Codon/ expression optim.
• Veneering high risk residues
• Surface charge heterogeneity
• Deamidation/ oxidation/ etc.
• Glycosylation prediction
• Metal cleavage sites
• Intein-like motifs
• Immunogenicity prediction
• etc., etc.
Portfolio Knowledge
Scientific literature
Rank Ordering
High Throughput BioChemical & BioPhysical Analysis (wet work)
•Inherent characteristics
Hydrophobicity, pl, etc.).
•Product Variants
Deamidation, oxidation, etc.).

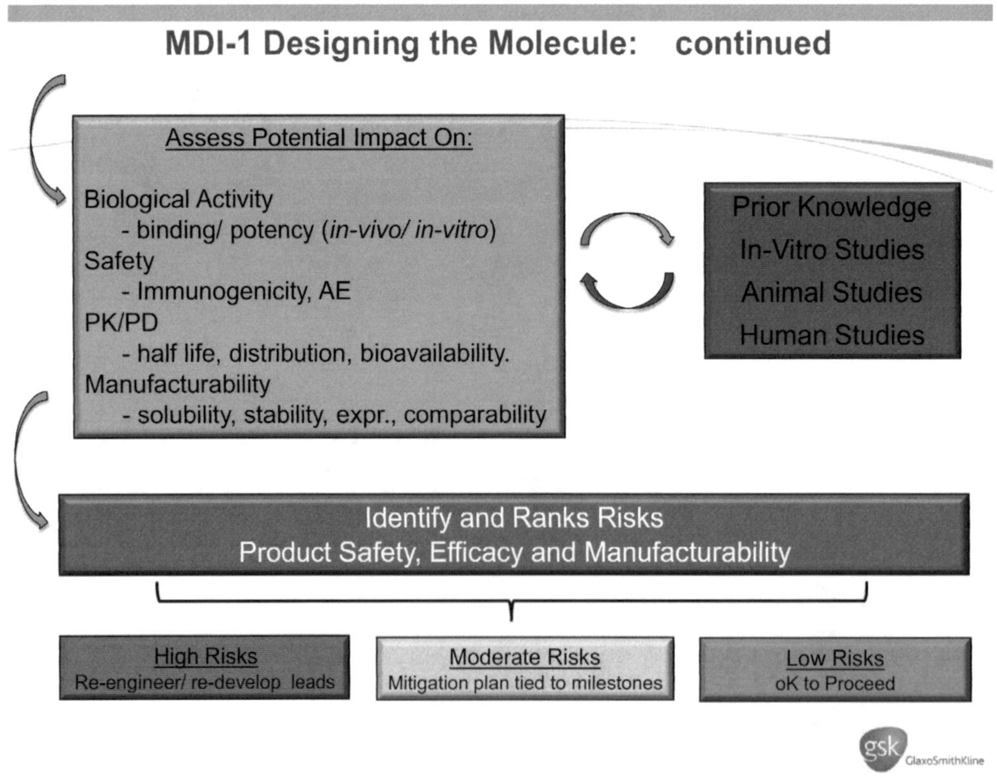
MDI-1 Designing the Molecule: continued
Assess Potential Impact On:
Biological Activity
- binding/ potency (in-vivo/ in-vitro)
Safety
- Immunogenicity, AE
PK/PD
- half life, distribution, bioavailability.
Manufacturability
- solubility, stability, expr., comparability
Prior Knowledge
In-Vitro Studies
Animal Studies
Human Studies
Identify and Ranks Risks
Product Safety, Efficacy and Manufacturability
High Risks
Re-engineer/ re-develop leads
Moderate Risks
Mitigation plan tied to milestones
Low Risks
oK to Proceed
gsk GlaxoSmithKline

## Molecular Design Intent 2: Developability Assessment

### Prospectively Defined – Pre-CMC engagement

- CMC/ biology risks
  - Development strategy (Fast to FTIH…Fast to Launch)
  - InVitro potency and its relation to biology/ MOA.
- Process targets
  - Purity, titer, yield, COGs sensitivity, etc.
- Formulation targets
  - Route, strength, shelf-life, device etc.
- Compatibility with established platforms
  - Process, analytical, formulation
- Early Clinical & Safety Demands
  - Quantity, strength, etc.

## MDI-2: Developability Assessment – continued

Scope modulated based on molecule/ developability assessment

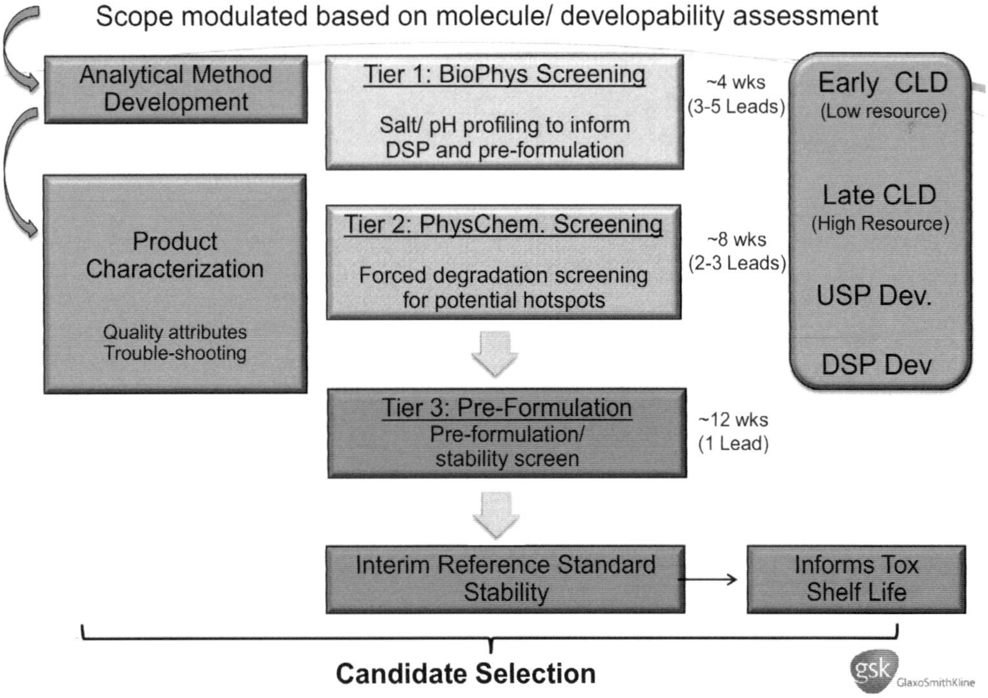

## Implementation of Technology Platforms

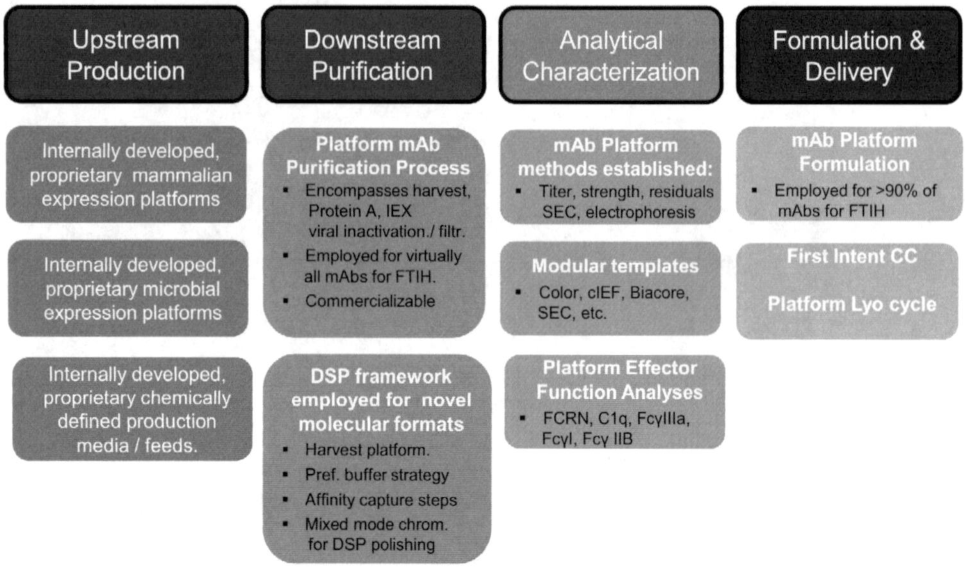


- Established technology platforms enable streamlining of development.
- Platforms harmonized in Discovery, Development, & Clinical Manufacturing

gsk GlaxoSmithKline

## Platform vs. Non-Platform Method Implementation

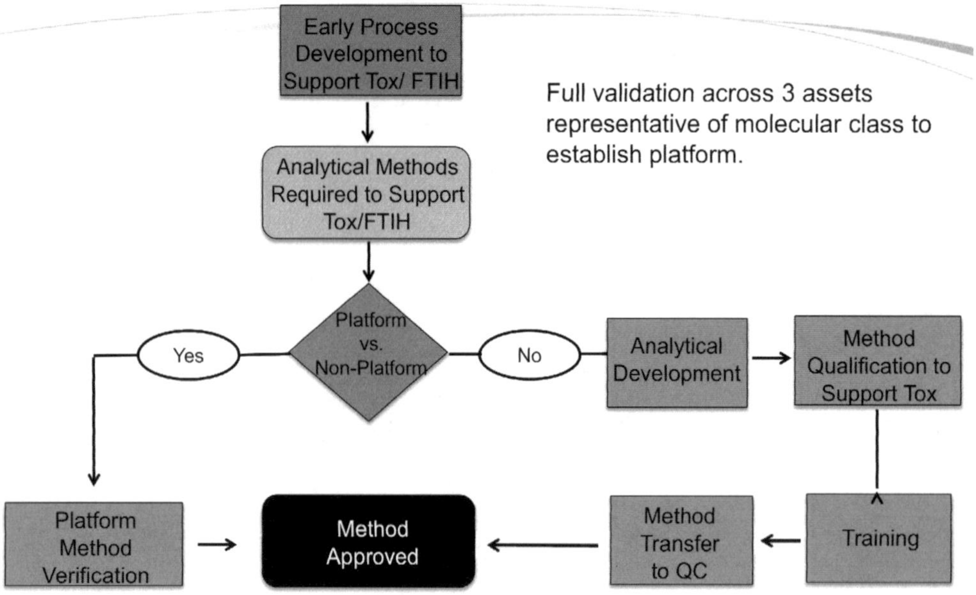


gsk GlaxoSmithKline

## Application of Molecular Design/ Platform Technologies

| mAb #1 Assumptions |
|---|
| • Predominant development risks with targeted biology<br>- No major CMC hurdles anticipated.<br>• Neutralization of soluble target<br>• No effector function involvement<br>• Moderate strength targeted w/IV infusion route. |

## Application of Molecular Design Intent & Platforms

### mAb Example #1

| | |
|---|---|
| Expression/USP Platforms | Reduced comparability risk |
| DSP Platforms | 2 chromatography step process (platform conditions). |
| Analytical Platforms | No effector function involvement/ platform analytics. |
| Formulation Platforms | Moderate strength targeted w/ IV infusion route (<100mg/mL) |

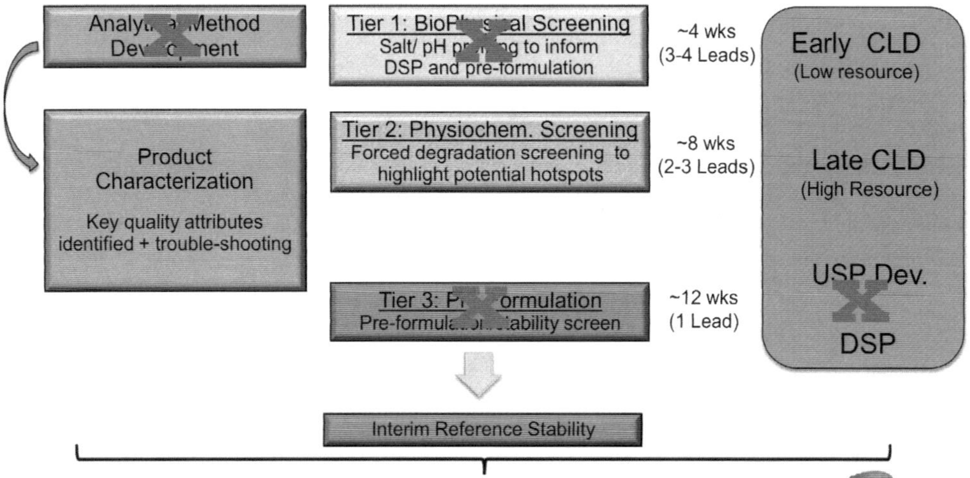


**Candidate Selection**

## Application of Molecular Design/ Platform Technologies

| mAb #2 Assumptions |
|---|
| • Significant clinical/commercial demand. |
| • Fc-effector function involvement anticipated. |
| • Minor propensity for aggregation |
| • High strength commercial Sub-Cut formulation |
| • Potential viscosity concern<br>-Computational ID of surface charge heterogeneity. |

## Application of Molecular Design Intent & Platforms

### mAb #2 Example

| | | |
|---|---|---|
| Expression/USP Platforms | | Increased comparability risk/ commercial demand or COGs pressure |
| DSP Platforms | | Minor aggregation |
| Analytical Platforms | | Poss. effector function involvement. |
| Formulation Platforms | | High strength Sub-Cut formulation, potential viscosity concerns |

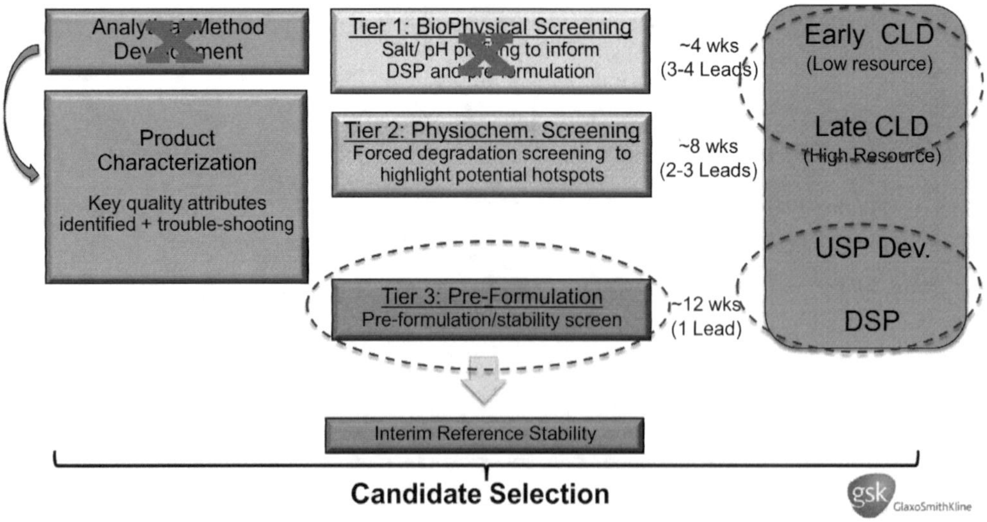

## Application of Molecular Design Intent & Platforms

### Timeline Implications – mAb A vs. mAb B

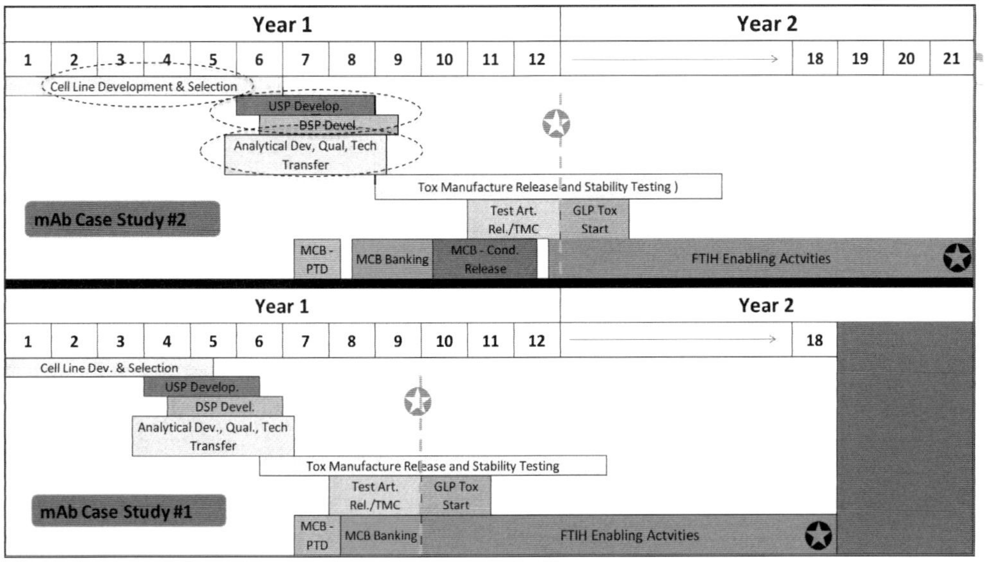


## Discussion Summary

- Formalised molecule & process de-risking approach
  - Ensures design of appropriate medicinal properties/ CMC targets.
  - Leverages *in silico* design & wet-work for candidate screening.
- Despite incorporating solid molecular design intent practices
  - Delivery of most mAbs achieved within 9-12 mos. (DNA to Tox).
    - Non-mAb formats ~9-16 mos.
- Insures Candidate Quality & risks are well understood at CS.
  - Optimum molecule selected.
  - Monitoring of portfolio quality risks at CS and beyond.
  - More rapid advancement to commercialization
  - Product attrition reduced during development lifecycle
  - Products that deliver greater value to a larger range of patients.

## Acknowledments


- BioPharm Discovery
  - Steve Martin
  - Mark Uden
  - Hella Bosteels
  - Gary Finka
  - James Walford
  - Victoria Higgins
  - Ahmed Yasin
  - Mike Molloy
- QbD
  - Aston Liu
  - Ilse Blumentals
- BioPharm Development
  - Narendra Bam
  - Yuan Zhu
  - Sybille Galosy
  - Prem Patel
  - Antonio Ubiera
  - Monica Branco
  - Doug Nesta
  - George Crotts
- GMP Ops/ QCC
  - Wayne Kelley
  - Michael Coppello

**Increasing Access to the Clinic with a Low Cost, Low Volume Manufacturing Process**

Salm, J., Pfizer, Inc

Originally presented on: 2/26/2013 14:15 - 14:45

As disease targets become increasingly complex, biological drug researchers are relying more heavily on studying diseases and mechanisms in humans. Development organizations are being asked to quickly manufacture safe, high quality clinical trial material to enable FIH studies. At the same time, more candidates are being investigated, which places a higher demand on limited development resources. To address this challenge, a new mAb development and manufacturing paradigm is proposed that eliminates process development and right-sizes the scale of manufacturing to match initial FIH study needs. Extensive platform knowledge and historical data were used to design the process so that it would be applicable to a broad subset of mAbs. The process relies on a pre-candidate selection screen designed to identify mAbs that are compatible with the proposed process. Proven manufacturing approaches make up the process to ensure quality and safety of the drug product, to minimize regulatory risk and to enable future program development: a clonal CHO cell line, a 12 day production bioreactor, a two column purification process, generic viral inactivation, a single drug product formulation and configuration and a single set of release assays. The resulting process reduces the cost and time to FIH by 50% and 35% respectively. In this talk, the proposed manufacturing process will be presented alongside the historical data used to develop it. The criteria for including a mAb in the process will be discussed along with how many mAbs are expected to meet the criteria. The quality of the product produced using the new process will be compared to material from a program involving traditional development. The impact of the process on late stage development and product lifecycle will also be considered.

**Accelerated Formulation Development for Antibody-Drug Conjugates (ADCs)**
Flores , H., Genentech
Originally presented on: 2/26/2013 14:45 - 15:15

Knowledge gained during development of the first antibody-drug conjugates (ADCs) to enter development has enabled the identification of a base or "platform" formulation for this new class of potential anti-cancer molecules. From this acquired knowledge, a formulation, Formulation A, has been selected which is suitable for use in pre-clinical toxicology studies. With only minor modifications for some ADCs, this platform formulation has also been shown to provide adequate shelf life to support early clinical trials. Additional challenges have resulted in a standardized approach to clinical administration. Utilizing this formulation development strategy, six additional ADCs have initiated human clinical trials within a period of two years.

AMGEN
Pioneering science delivers vital medicines™

# Development of Effective Control Strategies Using Product Quality Risk Assessment

Gregg Nyberg

Accelerating Biopharmaceutical Development and Quality by Design in Biologics

February 26, 2013, Coronado Island, CA

# An effective control strategy is a regulatory expectation and good business practice

- Guidance documents set expectations
  - Defined in ICH Q10 "A planned set of controls, derived from current product and process understanding, that assures process performance and product quality . . . "
  - ICH Q11: "Every drug substance manufacturing process . . . has an associated control strategy"
- Sound control strategies are good business practice
  - Ensure quality of our products
  - Provide focus for development and improvement efforts

2 AMGEN

## Control strategy integrates all aspects of process and product controls

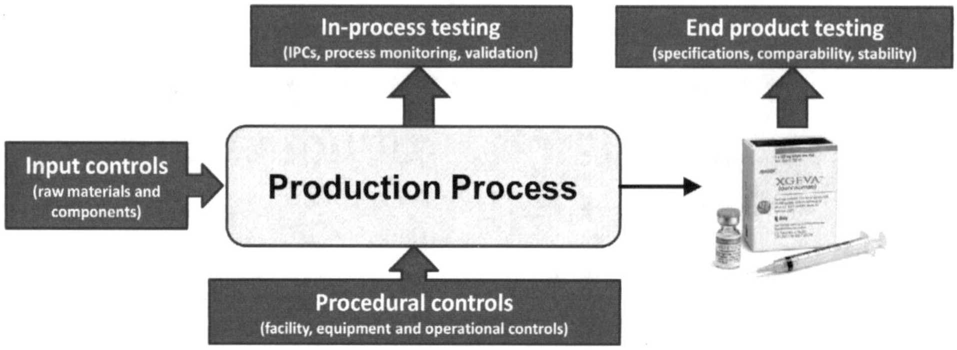


**Control elements are well coordinated and integrated in effective control strategies**

3

AMGEN

## Risk assessment is key for effective control strategy development

- Refinement and integration of control elements
- Prioritization of process development activities
- Assessment of changes and mitigation strategies
- Documentation of control strategy justification and rationale

**ICH Q10: "Use quality risk management to establish the control strategy"**

4

AMGEN

## Product quality risk assessment (PQRA) evaluates overall risk to patient

- Product Quality Risk Assessment
  - Considers all known Product Quality Attributes
  - Includes all steps leading up to administration to the patient
- Risk assessment inputs:
  - Comprehensive understanding of quality attributes [Severity]
  - Understanding of process impact on quality attributes [Occurrence]
  - Testing strategy/method capability [Detection]

Overall Risk = f ⟨ PQA Criticality *Severity* , Process Capability *Occurrence* , Testing Strategy *Detection* ⟩

5

## Product quality attribute (PQA) assessment determines criticality (severity)

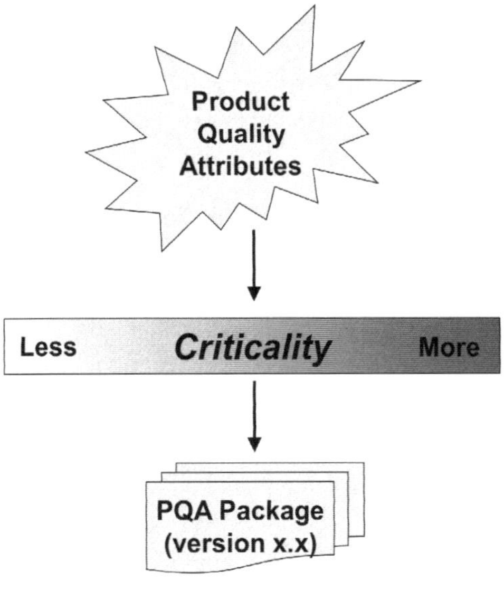


- Comprehensive assessment of all known quality attributes
- Foundation is systematic scientific evaluation of product attributes
- Severity scoring is based on safety and efficacy for product, platform and prior knowledge
- Effective knowledge management is key for state-of-the-art assessments

6

## Scoring of occurrence and detection is aligned with QbD concepts

- Occurrence scoring decision tree
  - Score reflects probability of attribute falling outside acceptable range
  - Higher score if no downstream redundancy / clearance capability

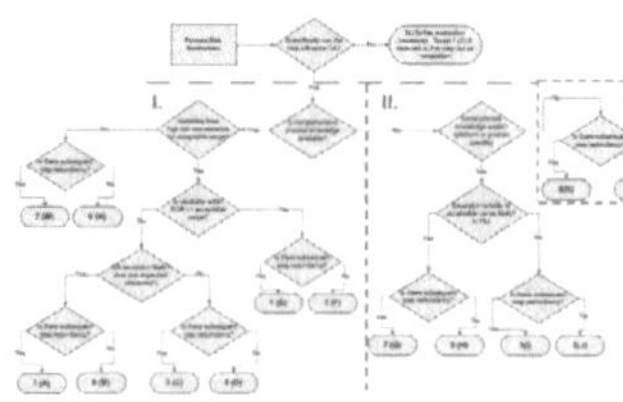

- Detection scoring combines two concepts
  - **Method capability** considers limit of quantitation, precision, specificity and orthogonality
  - **Control stringency** accounts for frequency of testing and limits applied

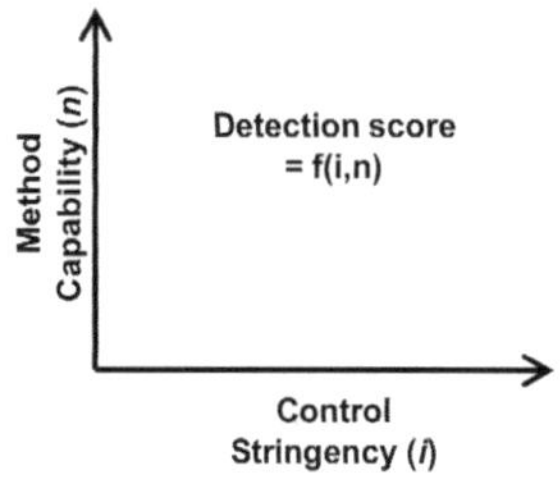


7

AMGEN

## Detection scoring takes into account method capability and control stringency

| *Detection Scoring* | | Control Stringency | | | | |
|---|---|---|---|---|---|---|
| | | Routine testing with reject limits i=1 | Routine testing with action limits i=3 | Routine monitoring i=5 | Periodic testing i=7 | Characterization i=9 |
| Method Capability | Qualitative n=9 | 5 | 6 | 7 | 8 | 9 |
| | Low precision, quantitative n=7 | 4 | 5 | 6 | 7 | 8 |
| | Not orthogonal, non-specific, precise n=5 | 3 | 4 | 5 | 6 | 7 |
| | Orthogonal, non-specific, precise n=3 | 2 | 3 | 4 | 5 | 6 |
| | Specific, precise n=1 | 1 | 2 | 3 | 4 | 5 |

8

AMGEN

# Severity and Occurrence scores define a Preliminary Hazard Risk Level

| | | | Severity of Risk | | | | |
|---|---|---|---|---|---|---|---|
| | | | Insignificant | Minor | Moderate | Major | Severe |
| | | | 1 | 3 | 5 | 7 | 9 |
| Likelihood of Occurrence | Frequent | 9 | Medium | Medium | High | High | High |
| | Likely | 7 | Low | Medium | High | High | High |
| | Occasional | 5 | Low | Medium | Medium | High | High |
| | Unlikely | 3 | Low | Low | Medium | Medium | High |
| | Remote | 1 | Low | Low | Low | Low | Medium |

- Highlights areas of process risk
- Inform process development and characterization
- Inform risk-based testing strategy



# Detection and Preliminary Hazard are combined to determine Overall Risk Level

| | | Detection | | | | | | | | |
|---|---|---|---|---|---|---|---|---|---|---|
| | | Almost Certain | Very High | High | Moderately High | Moderate | Slight | Remote | Very Remote | Absolutely Uncertain |
| | | 1 | 2 | 3 | 4 | 5 | 6 | 7 | 8 | 9 |
| Risk Classification (from Table 1) | High | Low | Medium | Medium | Medium | High | High | High | High | High |
| | Medium | Low | Low | Low | Low | Medium | Medium | High | High | High |
| | Low | Low | Low | Low | Low | Low | Low | Medium | Medium | Medium |

- Overall risk considers our ability to detect and control deviations to prevent impact to patients

# Assessment timing is optimized to support control strategy development

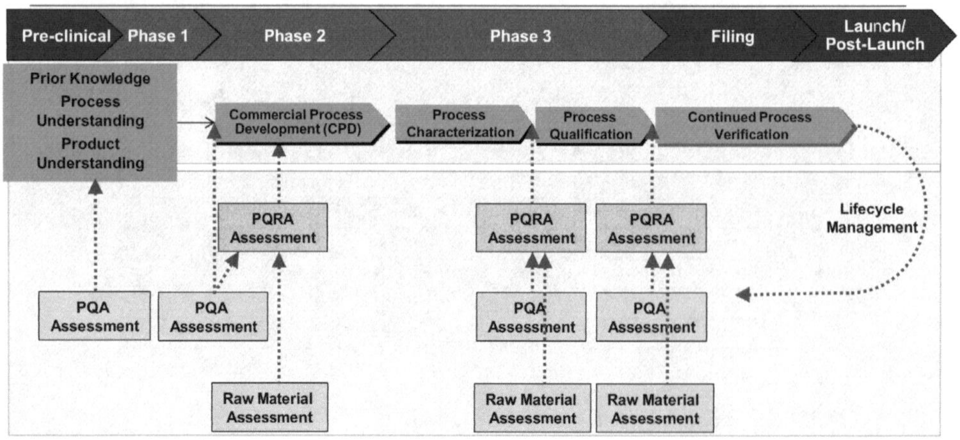




AMGEN

# Assessment timing is optimized to support control strategy development

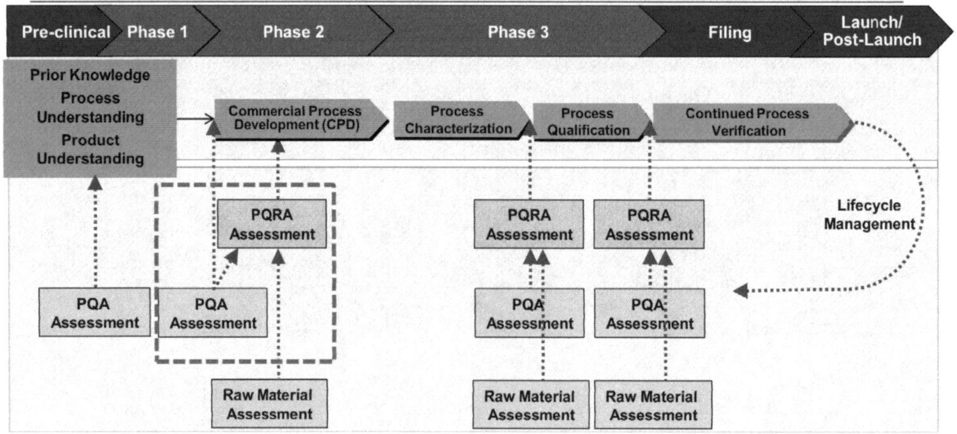


- PQRA scored at time of process lock
- Confirm proposed process has appropriate redundancy and robustness
- Make sure we are measuring the right things: identify analytical method development opportunities



AMGEN

## Assessment timing is optimized to support control strategy development

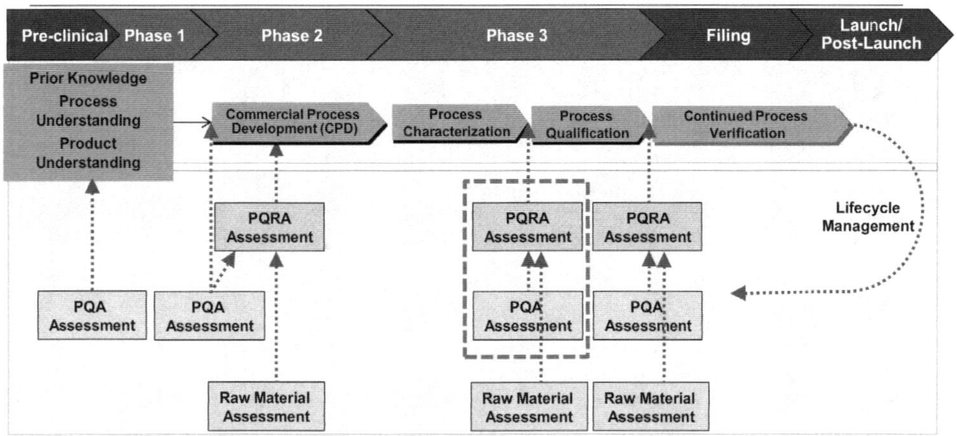


- Refine control strategy with improved process understanding
  - Define key/critical testing points
  - Finalize testing strategy for process qualification (performance qualification, comparability, in-process controls, specification)

13

AMGEN

## Assessment timing is optimized to support control strategy development

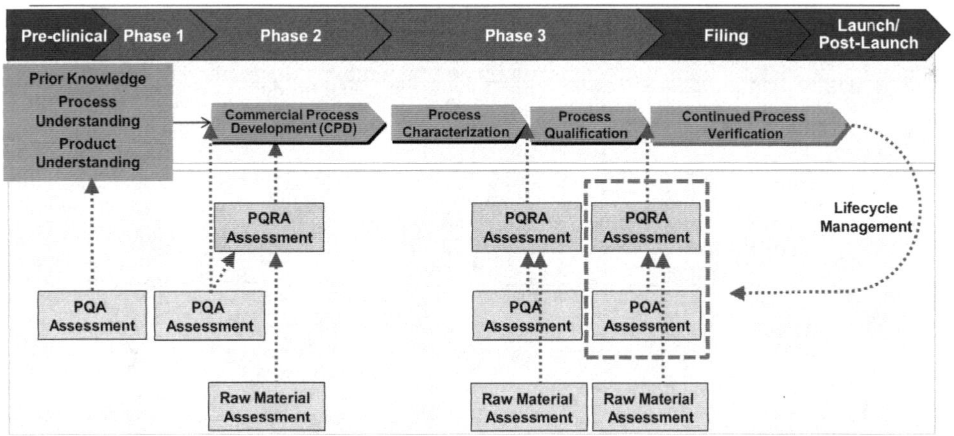


- Finalize control strategy
  - Establish test plan for routine manufacturing
  - Provide justification for specifications and in-process controls

14

AMGEN

# Post-approval lifecycle assessments may drive control strategy revisions

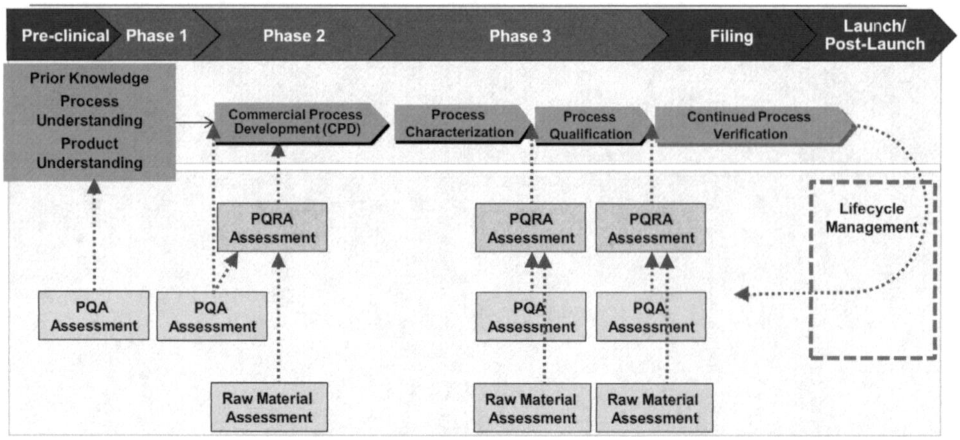


- **Ensure control strategy remains effective across product lifecycle**
  - **Evaluate impact of changes on control strategy effectiveness**
  - **Process and/or method capability signals can drive PQRA revisions and potentially control strategy changes**

15 AMGEN

# Example: CPD risk assessment identifies high risk

| Quality Attribute: | Potential Adverse Impact: | Severity |
|---|---|---|
| Attribute X | Attribute X impacts pharmacokinetics (increase in clearance) | 7 |

| Unit Operation | Correlation (↑,↓ or testing only) | Control Elements: Procedural Controls | Raw Materials and Components | Characterization Only | Process/Product Monitoring (FIO) | IPC (Action Limit) | IPC (Reject Limit) | Release Spec | Stability | Validation (PPQ, Transport, etc) | Comparability | Occurrence: Occurrence Supporting Information | Occurrence Decision Tree Code | Occurrence Score | Preliminary Hazard Risk Level | Detection: Detection Method | Capability (n) | Stringency (i) | Detection Score | Overall Unit Operation Risk Level |
|---|---|---|---|---|---|---|---|---|---|---|---|---|---|---|---|---|---|---|---|---|
| 6 Production Bioreactor | ↑ | | | | X | | | | | | | CPD data suggests excursions unlikely; operating parameter impact unknown; no downstream clearance | J | 5 | High | Detected by RP-HPLC as one of several species in peak B; routinely monitored in clinical mfg (FIO) | 5 | 5 | 5 | High |
| | Quality Attribute Overall Risk Level | | | | | | | | | | | | | | | | | | | High |

**Potential mitigation actions:**

- **Characterize/ improve process capability (occurrence)**
- **Add action/ investigation limit (detection stringency)**
- **Improve method specificity (detection capability)**
- **Establish clinical relevance (severity)**

16 AMGEN

# Example: control strategy refined following process characterization

| Quality Attribute: | Potential Adverse Impact: | Severity |
|---|---|---|
| Attribute X | Attribute X impacts pharmacokinetics (increase in clearance) | 7 |

| | | Control Elements | | | | | | | | | | Occurrence | | | | Detection | | | | |
|---|---|---|---|---|---|---|---|---|---|---|---|---|---|---|---|---|---|---|---|---|
| Unit Operation | Correlation (↑, ↓ or testing only) | Procedural Controls | Raw Materials and Components | Characterization Only | Process/Product Monitoring (FIO) | IPC (Action Limit) | IPC (Reject Limit) | Release Spec | Stability | Validation (PPQ, Transport, etc) | Comparability | Occurrence Supporting Information | Occurrence Decision Tree Code | Occurrence Score | Preliminary Hazard Risk Level | Detection Method | Capability (n) | Stringency (I) | Detection Score | Overall Unit Operation Risk Level |
| 6 Production Bioreactor | ↑ | | | | | | | | | | X | PC indicates minimal variability across operational space | F | 3 | Medium | Detected as single species by peptide map; comparability test to assess process/site changes | 1 | 7 | 4 | Low |
| | Quality Attribute Overall Risk Level | | | | | | | | | | | | | | | | | | | Low |

- **Highly capable process**
- **Periodic comparability testing to assess process/site changes**



AMGEN

# Example: process change drives changes to control strategy

| Quality Attribute: | Potential Adverse Impact: | Severity |
|---|---|---|
| Attribute X | Attribute X impacts pharmacokinetics (increase in clearance) | 7 |

| | | Control Elements | | | | | | | | | | Occurrence | | | | Detection | | | | |
|---|---|---|---|---|---|---|---|---|---|---|---|---|---|---|---|---|---|---|---|---|
| Unit Operation | Correlation (↑, ↓ or testing only) | Procedural Controls | Raw Materials and Components | Characterization Only | Process/Product Monitoring (FIO) | IPC (Action Limit) | IPC (Reject Limit) | Release Spec | Stability | Validation (PPQ, Transport, etc) | Comparability | Occurrence Supporting Information | Occurrence Decision Tree Code | Occurrence Score | Preliminary Hazard Risk Level | Detection Method | Capability (n) | Stringency (I) | Detection Score | Overall Unit Operation Risk Level |
| 6 Production Bioreactor | ↑ | X | | | | X | | | | X | X | Optimized process requires control of parameter Z to maintain acceptable Attribute X | D | 5 | High | Detected as single species by peptide map; validated with process change, IPC with action limit | 1 | 3 | 2 | Medium |
| | Quality Attribute Overall Risk Level | | | | | | | | | | | | | | | | | | | Medium |

- **Change in process capability drives control strategy change**
- **Procedural and detection controls leveraged to maintain acceptable risk profile**



AMGEN

## PQRAs have provided value to teams

- More focused process and method development
  - Prioritized process development/characterization activities
  - Streamlined identification of sample types for method development / qualification
- Improved technology transfer
  - Identified gaps in process/product understanding
  - More focused discussions (internal and with third party partners)
- Improved control strategy development
  - Refined comparability test plan
  - Reduced cost of quality by eliminating redundant, non-value added testing

19 AMGEN

## Conclusions

- A Product Quality Risk Assessment tool has been generated that applies Quality by Design principles
- The risk assessment tool has been integrated into our business practices
- The PQRA has been a useful tool for project teams

20 AMGEN

# Acknowledgments

QbD Control Strategy Core Team
- Gregg Nyberg
- Cenk Undey
- Melissa Weakly
- Rob Woolfenden
- Chris Acklin
- Tom Monica
- Paul Konold
- Linda Lai
- Paul Konefal
- Claudia Baikalov
- Beatriz Mesa

PQRA Working Team
- Karen Schlobohm
- Kathy Leach
- Stan Shimizu
- Nolan Polson
- Michelle Priem
- Athena Nagi
- Carlos Monteagudo

Quality by Design Team
- Bob Kuhn
- Greg Flynn
- Dean Petit
- Carole Heath
- Ed Walls
- Drew Kelner
- Paul Tsang



# Questions?

**The Appropriate Use of Risk Management Tools in Biopharmaceutical Development**

Subashi , A., Pfizer

Originally presented on: 2/26/2013 16:50 - 17:20

The talk will cover the appropriate use of risk management tools in biopharmaceutical development. Included will be a discussion of use of quality risk management approaches to assess attribute criticality, process parameter assessments and the control strategy. This will also include feedback gained through the QbD pilot and other agency interactions.

**The Development of a Rational Testing Strategy using QbD: A Case Study**

Cromwell , M., Genentech

Originally presented on: 2/26/2013 17:20 - 17:50

A control strategy is required for every product to ensure product quality. The elements that comprise the control strategy include controls associated with the environment, procedures, incoming raw materials, process parameters, and testing. The use of risk assessment tools facilitates the development of the control strategy, particularly with respect to parametric and testing control. A case study will be presented in which a risk ranking filter was established to guide the development of a rational testing strategy for a liquid monoclonal antibody product. Examples of the application of the tool for specific CQAs will be discussed along with health authority feedback regarding the resulting testing strategy.

Practical considerations to risk assessments for efficient use in pharmaceutical development and production

Dr. Christoph Stark

Feb. 26, 2013

NOVARTIS

## Agenda

I) Problem setting: Factors influencing Risk Assessment in pharmaceutical industry

II) Novartis approach to Risk Assessment

III) The "Universe of Risks" – the world beyond process parameters

## Factors influencing Risk Assessment of drug product formulation and process

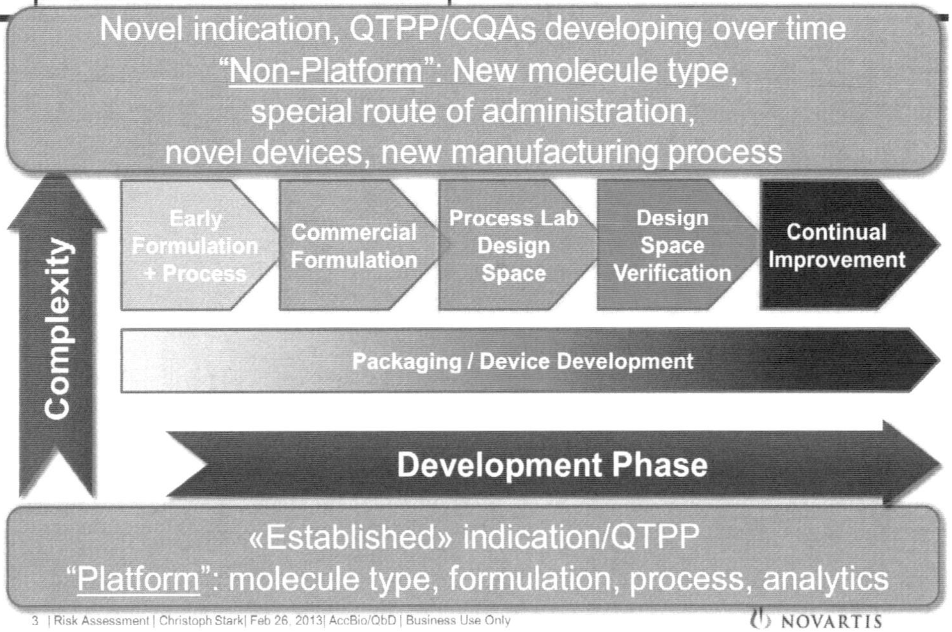




## ...and do not forget about the third factor .....

People

- Early phase, late phase, commercial
- Formulators, process developer, engineer, packaging developer, device developer, production line operators, MS&T, analysts, QC, technical project management
- CROs, CMOs
- QA
- Management
- Health Authorities

## Risk assessment tools

### ICH Q9:

- Basic risk management facilitation methods (flowcharts, check sheets etc.);
- Failure Mode Effects Analysis (FMEA);
- Failure Mode, Effects and Criticality Analysis (FMECA);
- Fault Tree Analysis (FTA);
- Hazard Analysis and Critical Control Points (HACCP);
- Hazard Operability Analysis (HAZOP);
- Preliminary Hazard Analysis (PHA);
- Risk ranking and filtering;
- Supporting statistical tools.

### Others:

- PRA, RCA, HRA, CCFA, .................many more.



## RA in Process development

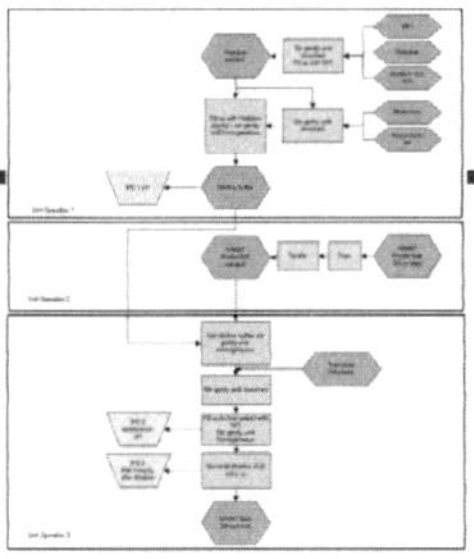

1. **Ishikawa diagram / Fault tree analysis**
2. **FMEA:**

**RPN = Impact x Occurrence x Detectability**

- Modular RA with generic risk assessment chapters (= unit operations)
- Only project specific Failure Modes are evaluated each time
- Selection of Failure Modes to be assessed done in fishbone/ FTA, based upon platform knowledge and process data available
- For a «Platform» molecule/process this should be a pretty straightforward exercise

# The first Risk Assessment for a “Non-Platform” molecule/process (1)

**Basis prior start of process development**

- Some understanding on «Impact»
- An educated guess on «Occurrence»
- Very limited idea on «Detectability» as control strategy not yet established

By experience interpretation of «Detectability» varies significantly between different scientific disciplines (e.g. engineer vs. analyst)

**Performing a full FMEA at this stage has the risk to:**

- Rate preliminary CPP/CMA too low due to a perceived tight control strategy which is later on lost -> missed CPP w/o adequate risk assessment and management
- Fix early on an unnecessarily extensive control strategy which has wrong focus = distraction from critical parameters = risk to the process in general



# The first Risk Assessment for a “Non-Platform” molecule/process (2)

Stepwise approach:

1. Determine **Criticality = Impact x Occurrence** = Preliminary Hazard Analysis (PHA)
2. Do experiments and **collect database**
3. Set up accordingly first version of **Control Strategy**
4. Update RA by now adding «Detectability»
5. Determine **RPN = Impact x Occurrence x Detectability**

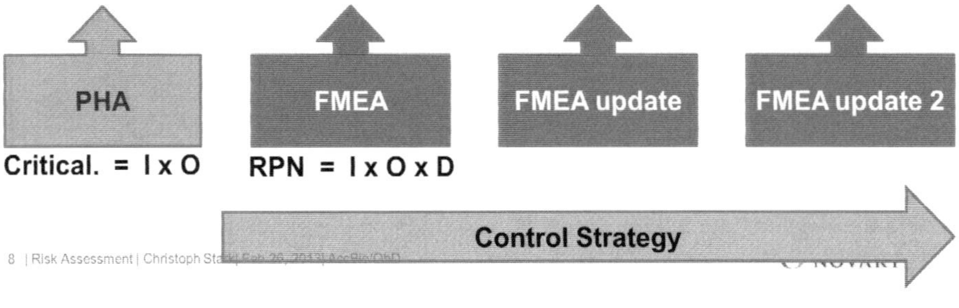

## RA in process development "Platform" vs. "Non-platform" molecule/process

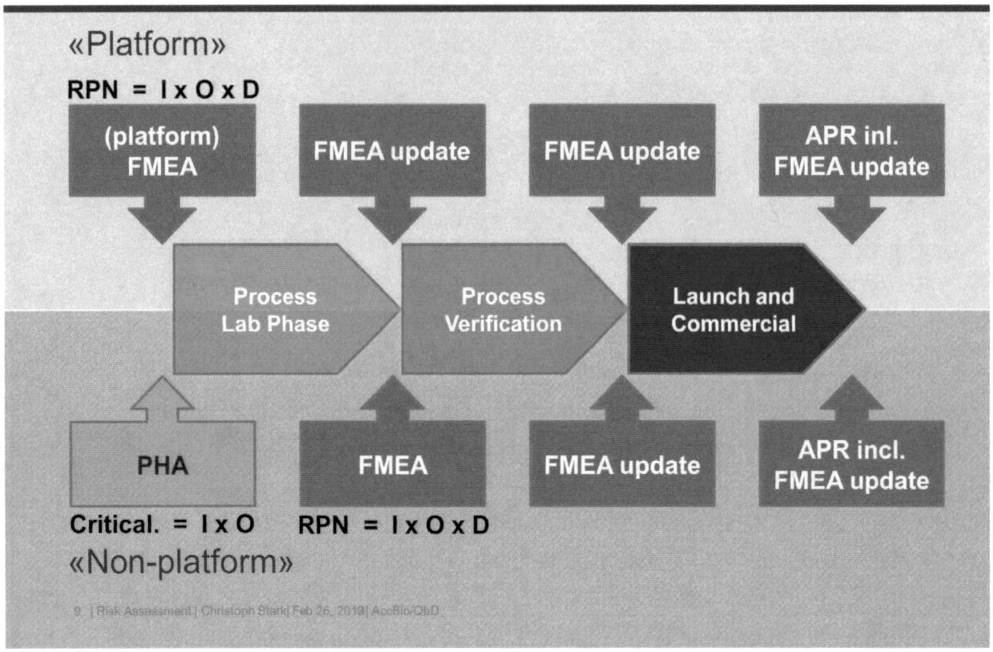


## «Non-platform», PHA transitions into FMEA

| No. | Process Step | Related process parameter or variable | Deviation of variable or parameter or potential failure | Potential Cause(s) | Previous evaluation | Effect on product quality and process observed during previous evaluation | Reference | Probability | Potential Impact on Product Quality | Impact | Current Detection Mode or Control Mechanism | Detectability | Risk Priority Number |
|---|---|---|---|---|---|---|---|---|---|---|---|---|---|
| 3.17 | Sterile filtration | Filter pressure | Too high pressure | Operator error | none | early molecule characterization | Report X | 5 | Aggregate formation | 9 | - Adjustment with manometer<br>- Documentation pressure in BR<br>- Release analytics | 2 | 90 |

- **Same scoring tables** used in PHA and FMEA for assessment of Impact and Occurrence
- **Different RPN thresholds** for PHA vs. FMEA



NOVARTIS

## Scoring tables – how to achieve consistency over time

- Do not change scoring tables between development and commercial
  - Consider need for fine graduation for commercial products when lots of process performance data available
  - If helpful, limit choice in development phase to few selected levels within given scoring table

- Be quantitative rather than qualitative in description
  - E.g. use «0-20% change in bioactivity or PK" rather than "some degree of customer dissatisfaction"



NOVARTIS

## Linking studies – Risk Ranking and Filtering

| Formulation parameter | Component | Main effect score | Rationale | Interaction score | Potential interaction parameters | Severity | Recommended study |
|---|---|---|---|---|---|---|---|
| pH | Buffer | **8 = high** | AP in SEC at higher pH | **5 = unknown** | Stabilizer, API | 40 | Multivariate |
| Buffer | Buffer salt | **2 = low** | No effect seen in Stability program | **1 = no** | API | 2 | No additional study |

DoE



NOVARTIS

# THE «UNIVERSE OF RISKS» IN DRUG PRODUCT DEVELOPMENT

# 3 CASE STUDIES



## Material Attribute «excipient quality» (Case 1)

- Product 1 contains Polysorbate 20
- Pharmacopeial (Ph.Eur/USP) quality is not sufficient to ensure optimal product stability
- Individual control strategy needed

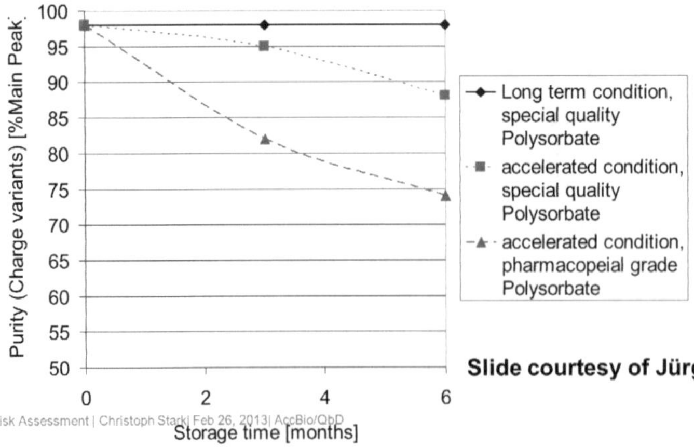


**Slide courtesy of Jürgen Sigg**

## Material Attribute «primary container quality» (Case 1)

- Leachables from different container qualities might significantly affect stability of the commercial formulation

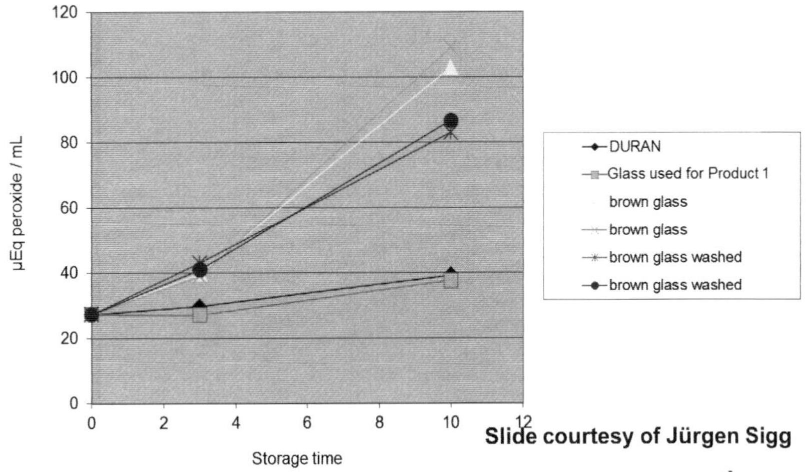


**Slide courtesy of Jürgen Sigg**



NOVARTIS

## Material Attribute «primary container quality» (Case 2)

*Quality attributes of pre-fillable syringes evaluated*

| | | Manufacturer 1 | | Manufacturer 2 | | Manufacturer 3 | | Manuf. 4 |
|---|---|---|---|---|---|---|---|---|
| | | Quality 1 | Quality 2 | Quality 1 | Quality 2 | Quality 1 | Quality 2 | Quality 1 |
| PDMS [mg/barrel] | Spec. | 0.8 ± 0.4 | 0.4 ± 0.1 | 0.5 ± 0.2 | 0.3 ± 0.1 | 0.8 ± 0.2 | *n.a.* | 0.55 ± 0.25 |
| | actual | 0.32 | 0.37 | 0.34[2)] | 0.48 | 0.18 | *n.a.* | 0.46 |
| Particles ≥ 10 μm/mL[1)] | | 1'422 | 955 | 7'817[2)] | 14'727 | 1'417 | *n.a.* | 17'730 |
| Turbidity[NTU] [1)] | | 1.8 | 1.6 | 8.9[2)] | 9.7 | 5.4 | *n.a.* | 9.5 |
| PDMS in DP 6M/5°C [ppm] [3)] | | 0.9[4)] | 0.3 | 0.7 | 1.7 | 1.1 | 1.0 | 1.8 |
| PDMS in DP 6M/40°C [ppm] [3)] | | 3.7[5)] | 7.6 | 19.5 | 7.5 | 2.1 | 6.2 | 19.5 |

*PDMS: Polydimethylsiloxan, [1)]in 0.04% PS80 solution after 3 months storage at 40 C,* [2)] results from same reference but different lot, [3)] *as assessed by AAS,* [5)] *level determined in 0.5mL fill, level determined in 1.0mL fill: 0.6 ppm after 5 C, 1.1 ppm after 25 C storage*

**Slide courtesy of Kathrin Schersch**



NOVARTIS

# Material Attribute «pimary container quality» (Case 2)

*Quality attributes of pre-fillable syringes evaluated*

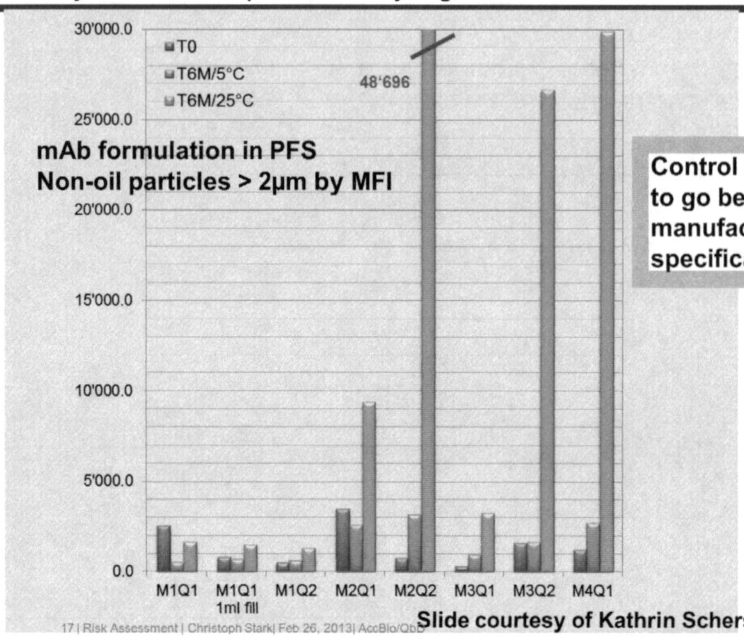


**Control strategy needs to go beyond manufacturer specifications**

**Slide courtesy of Kathrin Schersch**

NOVARTIS

# Linking DS process to DP process

*Filter blocking in DP manufacture after DS site change*

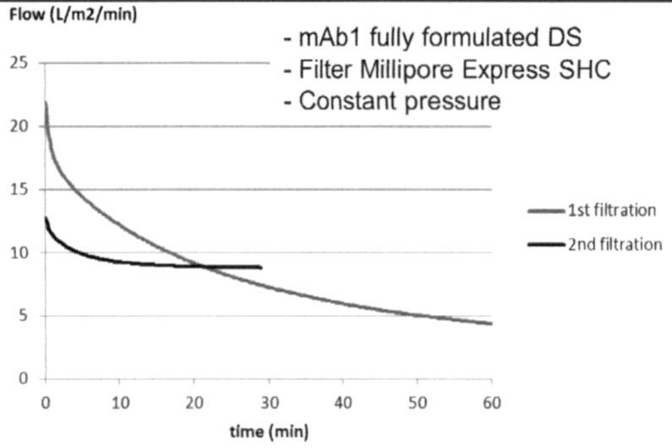


→ Pronounced filter fouling for mAb1 first filtration despite <u>comparable analytical characteristics prior first and prior second filtration</u>

**mAb1 prior 1st filtration**
NTA data below LOQ
(< $10^7$ particles per mL)

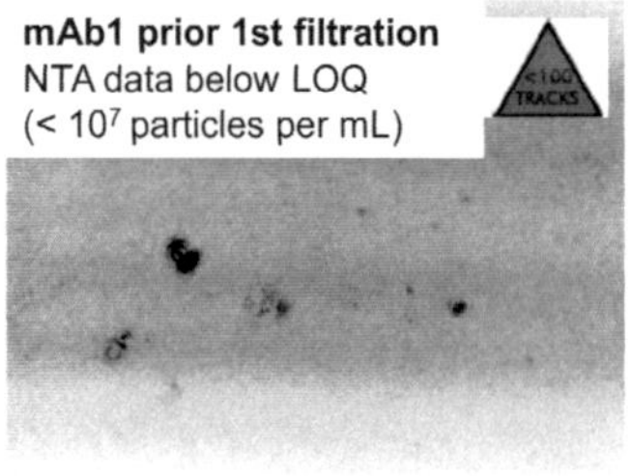

Sample Video Frame

| | SEC AP (%) | Light Obscuration (number of particles / mL) 2-5 µm | 10-25 µm | > 25 µm | DLS Pd (%) | Turb (NTU) |
|---|---|---|---|---|---|---|
| **mAb1 prior 1st filtration** | 0.2 | 2599 | 158 | 10 | 47 | 5 |
| **mAb1 prior 2nd filtration** | 0.3 | 3150 | 126 | 11 | 42 | 4 |

**Data: A. Bourgoin, M. Cosenza, M.Giffard. B. Ossuli, V. Rombach**

NOVARTIS

## Linking DS process to DP process (2)

*Filter blocking in DP manufacture after DS site change*

Most probable Root Cause found for membrane fouling was changed stirring mode at DS compounding which was not recognized as critical because DS control strategy did not capture submicron particles generated. Investigations currently finalized.



## Conclusion

- Use limited number of RA tools
- Consistently use same tool(s) and scoring tables over entire product lifetime
- Standardize format and scoring tables
- Use quantitative description in scoring tables rather than qualitative
- Establish modular approaches and generic modules as far as appropriate
- Focus efforts on molecule and process individualities
- For first RA do not work with assumed control strategy, rather leave out «Detectability» first and do PHA, transition to FMEA later
- Consider the «universe of risks», esp. the Material Attributes e.g. excipient and primary packaging quality

# Acknowledgements

- Karin Schönhammer
- Sherry Martin-Moe
- Kamal Egodage
- Jürgen Sigg
- Susanne Jörg
- Kathrin Schersch
- Michael Boychyn
- Andreas Fisch
- Marion Giffard



NOVARTIS

# Questions



NOVARTIS

MOMENTA

# Complexity in Biological Products – Applications to QbD

*Ganesh V Kaundinya, PhD*

*SBE's 3rd International Conference on Accelerating Biopharmaceutical Development*
*February 24 – 27, 2013*

## What is Complexity?

- Not Simple
- Not the same as complicated
- Not random or chaotic

- Phenomenon emerging from a collection of interacting objects
- Latin -- entwined or twisted together
- No simple Cause-Effect relationship

## A continuous spectrum

**Complicated**
Many Variables
Mathematically definable
Cause-effect mapped

**Chaotic**
Many Variables
"randomness" in behavior
Unpredictable

**Simple**
Few Variables
Linear response

**Complex**
Large number of Variables
Mathematically definable
Non-linear cause-effect – indeterminable
Can be "rules" based



## Simple: Car Key

**Simple** Complicated Complex Chaotic

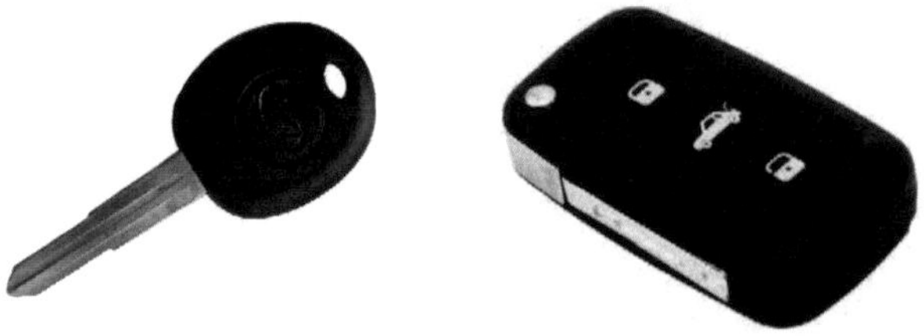

## Complicated: Car

Simple | **Complicated** | Complex | Chaotic

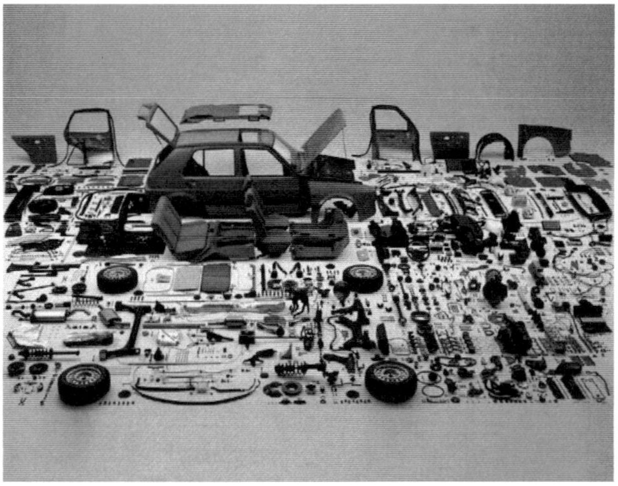

5 MOMENTA

## Complex: Driving in a City

Simple | Complicated | **Complex** | Chaotic

6 MOMENTA

## Elements of Complexity

1) Complex systems are made up a large number of components.
2) Web of connections between the parts, which leads to coupling and feedback.
3) The components are interdependent, i.e. removing a piece changes the function of the remainder.
4) Complex systems are open to being affected by events outside their boundaries.
5) They display synergy, i.e. the combined effect of changes to individual components differs in kind from the sum of the individual changes.
6) They exhibit non-linear behavior, in that a change in a system can produce an effect that disproportional to the cause.
7) Outliers contain significant information about the system behaviour – traditional statistical methods can be misleading

- *Adapted from http://deepfreeze9.blogspot.com/2007/09/software-complex-vs-complicated.html*
  *Homer-Dixon, "the ingenuity gap"*

7 MOMENTA

## Complicated

- Watch has many parts each well characterized and when combined will work in a deterministic way

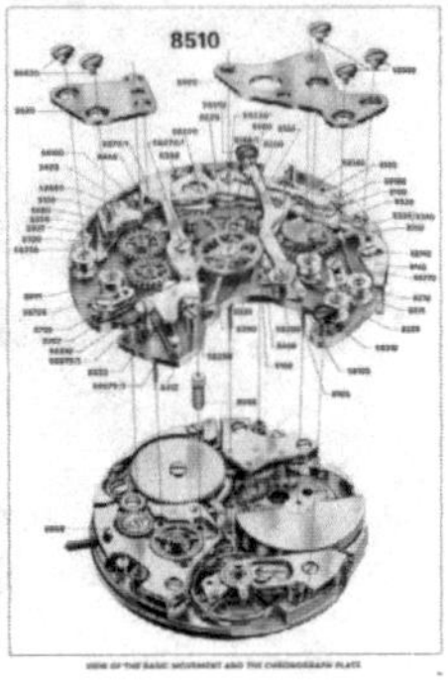

Precise Characterization of components *ESSENTIAL and SUFFICIENT* to understand, predict and control system behavior

8 MOMENTA

## vs. Complex System

Precise Characterization of components *is Necessary but not Sufficient* to understand and control system behavior



## Are biological systems complex?

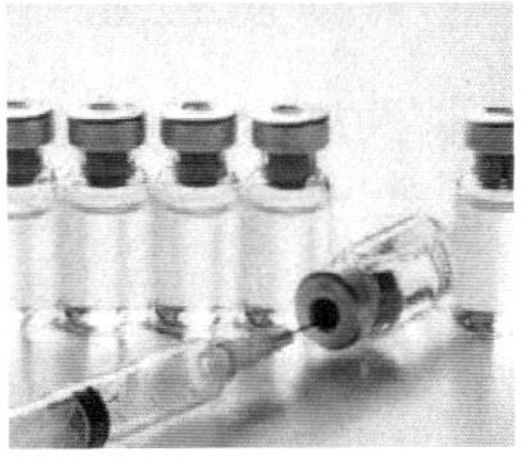

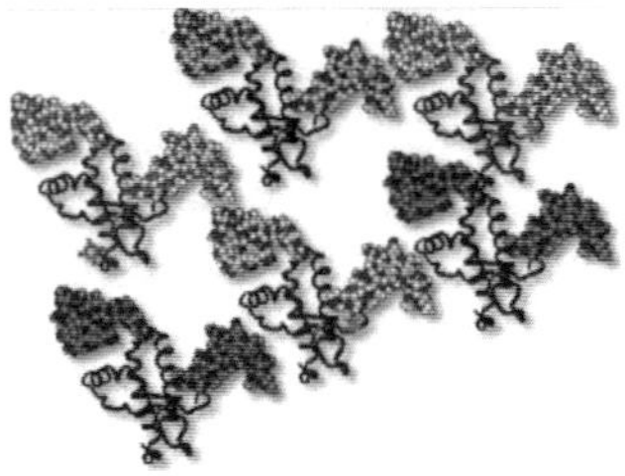

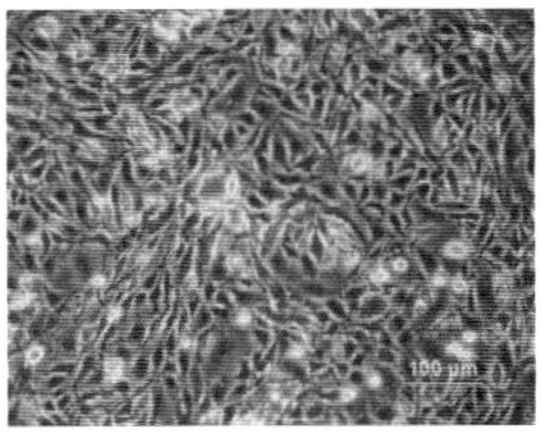

- Human Biology?
- Biologic Product (e.g. mAb) with micro-heterogeneity?
- Cells in a bioreactor?

## How do we think about complexity?

Symmetry – structured – ordered (Crystal)

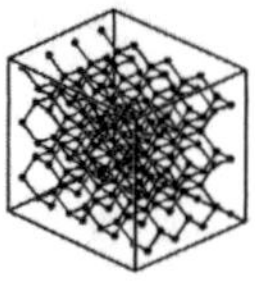

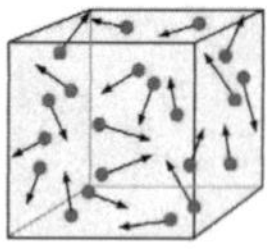

Chaotic – random – statistical distribution (gases)

**Complexity**

Network of interacting elements with "entropic" motions about constrained "nodes"

Key "signatures" that characterizes the system (aka. fingerprints)
Understanding of parameters that influence the fingerprints
Control



## Everyday examples

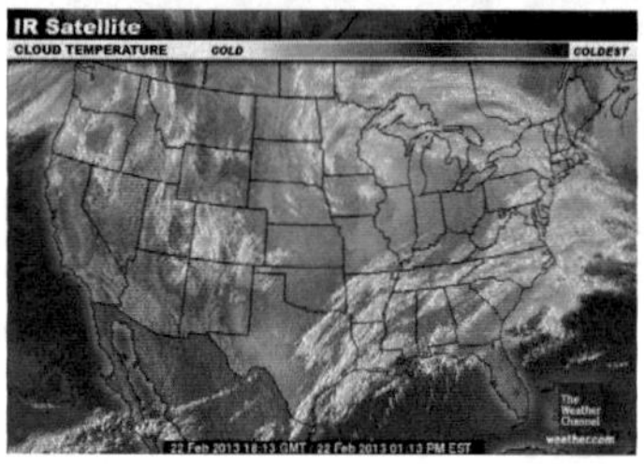


| | PECOTA | | | | ACTUAL | | | | |
|---|---|---|---|---|---|---|---|---|---|
| | AVG | OBP | SLG | OPS | AVG | OBP | SLG | OPS | OPS Diff |
| Raburn | .256 | .330 | .459 | .789 | .291 | .359 | .533 | .892 | 0.103 |
| Thomas | .219 | .292 | .346 | .638 | .240 | .324 | .385 | .709 | 0.071 |
| Santiago | .243 | .307 | .343 | .650 | .267 | .318 | .386 | .704 | 0.054 |
| Cabrera | .296 | .371 | .530 | .901 | .324 | .396 | .547 | .943 | 0.042 |
| Polanco | .292 | .334 | .385 | .719 | .285 | .331 | .396 | .727 | 0.008 |
| Everett | .233 | .283 | .325 | .608 | .238 | .288 | .325 | .613 | 0.005 |
| Thames | .234 | .304 | .486 | .790 | .252 | .323 | .453 | .776 | -0.014 |
| Inge | .244 | .321 | .415 | .736 | .230 | .314 | .406 | .720 | -0.016 |
| Guillen | .271 | .349 | .427 | .776 | .242 | .339 | .419 | .758 | -0.018 |
| Ordonez | .297 | .361 | .470 | .831 | .310 | .376 | .428 | .804 | -0.027 |
| Granderson | .266 | .342 | .467 | .809 | .249 | .327 | .453 | .780 | -0.029 |
| Laird | .253 | .306 | .399 | .705 | .225 | .306 | .320 | .626 | -0.079 |

# MOMENTA

## Example 1 – Enoxaparin ™

## Example 1 – Enoxaparin Sodium: Analysis of Chemical Complexity

Enoxaparin Sodium is a complex mixture low molecular weight heparin (LMWH) anticoagulant product.

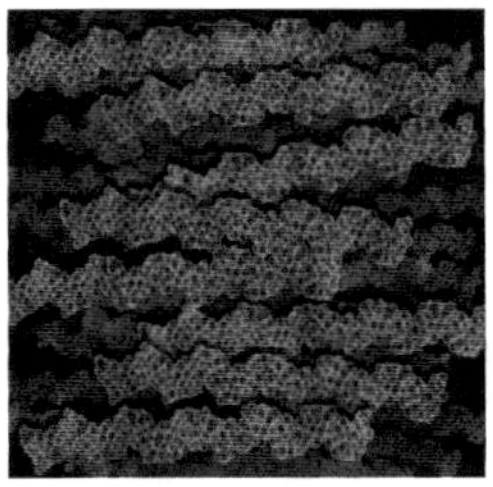

Extensive Knowledge on chemistry and biosynthesis

~ 1000 specific metrics (~ 50 methods)

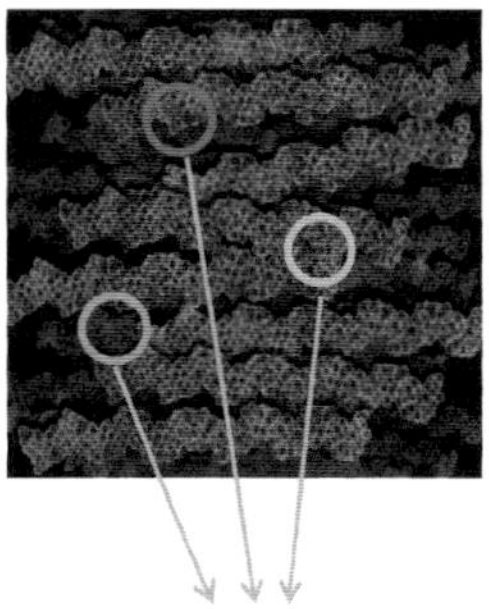

Enoxaparin Sodium (Thousands of chains, varied lengths, over $10^9$ structural possibilities)

Identification of key chemical fingerprints related to ***product-process understanding***

Developed an equivalent complex LMWH product as an interchangeable generic to the marketed enoxaparin product (Lovenox®).

# M-Enoxaparin – Generic Lovenox®
## *Approved & Launched July 23, 2010*

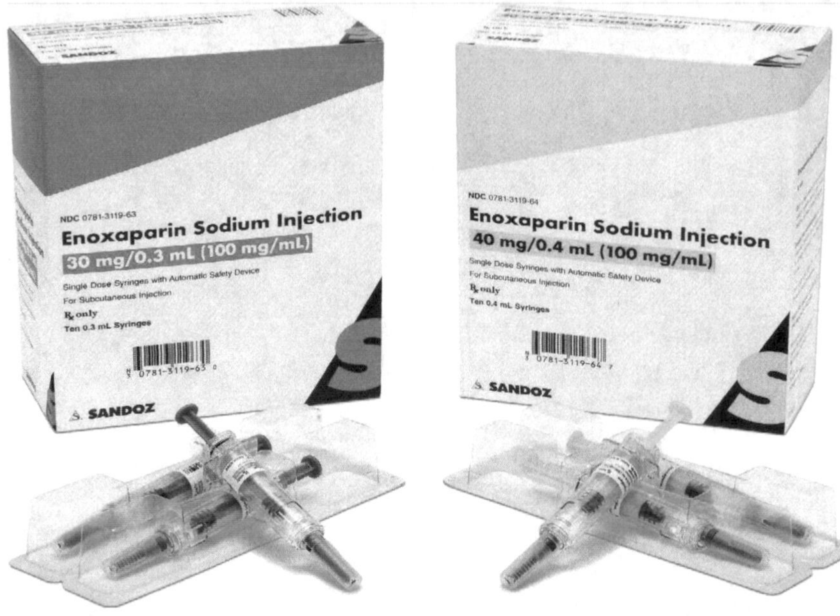


15 MOMENTA

# Complexity captured in the "5 Point criteria"

DEPARTMENT OF HEALTH & HUMAN SERVICES

Food and Drug Administration
Rockville MD 20857

JUL 23 2010

**III. GENERIC ENOXAPARIN CAN CONTAIN THE SAME ACTIVE INGREDIENT (ENOXAPARIN) AS LOVENOX WITHIN THE MEANING OF THE ACT AND FDA REGULATIONS**

As discussed above, an ANDA applicant for generic enoxaparin must provide sufficient information to show that the generic drug product contains the "same" active ingredient (enoxaparin) as Lovenox. We have concluded, based on our evaluation of current data and other current relevant scientific information and our scientific experience and expertise, that the following five criteria (or standards for identity) together provide sufficient information to conclude that generic enoxaparin has the "same" active ingredient as Lovenox:[48]

1. Equivalence of physicochemical properties
2. Equivalence of heparin source material and mode of depolymerization
3. Equivalence in disaccharide building blocks, fragment mapping, and sequence of oligosaccharide species
4. Equivalence in biological and biochemical assays
5. Equivalence of in vivo pharmacodynamic profile

16 MOMENTA

# "Fingerprints" provide insight into process

Contents lists available at ScienceDirect

Carbohydrate Polymers

journal homepage: www.elsevier.com/locate/carbpol

Identification of a novel structure in heparin generated by potassium permanganate oxidation

Daniela Beccati[a,1], Sucharita Roy[a,1], Fei Yu[a], Nur Sibel Gunay[a], Ishan Capila[a], Miroslaw Lech[a], Robert J. Linhardt[b], Ganesh Venkataraman[a,*]

[a] Momenta Pharmaceuticals, 675 West Kendall Street, Cambridge, MA 02142, USA
[b] Departments of Chemistry and Chemical Biology, Biology, and Chemical and Biological Engineering, Rensselaer Polytechnic Institute, Center for Biotechnology and Interdisciplinary Studies, 110 8th Street, Troy, NY 12180, USA

- Oxidation with permanganate results in the conversion of *N*-acetyl glucosamine at the reducing end of heparin chains to an *N*-acetyl glucosaminic acid which yields a characteristic signal at 2.10 ppm in the $^1$H-NMR spectrum of the heparin.



# Control Strategy

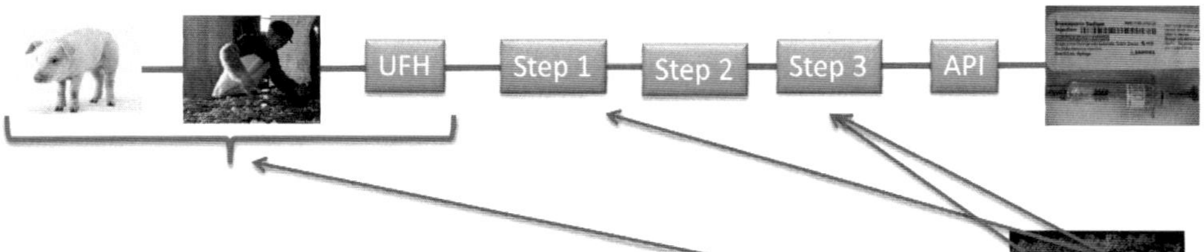


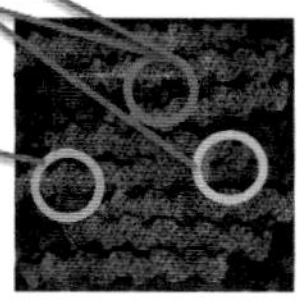

- Map the product-Process relationship
- Control strategy based on source and modification of the fingerprints
  - Certain fingerprints controlled as UFH signatures – implications to supply chain
  - Certain fingerprints created/modified during processing steps –> in process control

MOMENTA

## Example 2 – Copaxone™

## Copaxone ™

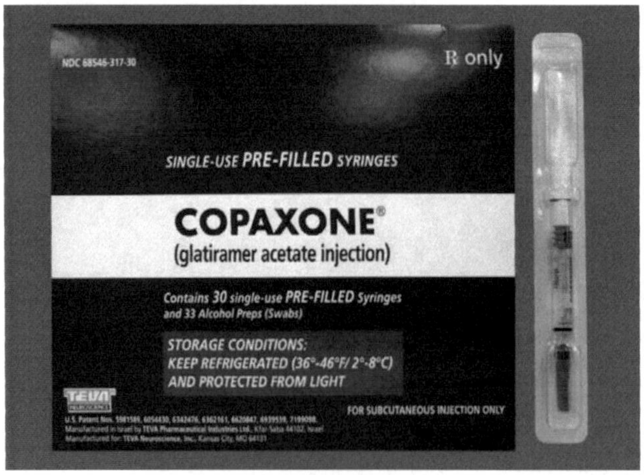


20 MOMENTA

## Copaxone® (glatiramer acetate) Product Description

- Acetate salts of synthetic polypeptides, containing four naturally occurring amino acids combined in specific ratios:
  - L-glutamic acid
  - L-lysine
  - L-alanine
  - L-tyrosine
- Molecular Weight = 5,000 – 9,000 daltons

MOMENTA

## Example 2 – Glatiramer Acetate: Managing Chemical and Biological Complexity

**Chemical Complexity**: Glatiramer Acetate is a complex peptide mixture. ( ~ $10^{29}$ theoretical chain structures spanning varied lengths).

+

**Biological Complexity**: Glatiramer Acetate is used for treatment of relapsing remitting multiple sclerosis. Mechanism of action in debate.

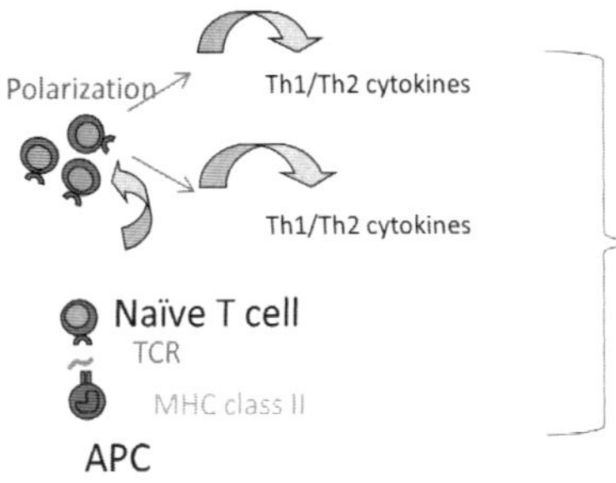


Understanding of critical ***nodes*** related To Glatiramer Acetate biology

Characterization and equivalence

22 MOMENTA

MOMENTA

## Example 3 - Biosimilars

## Momenta's FOB Strategy: Not a Similar; a True Generic

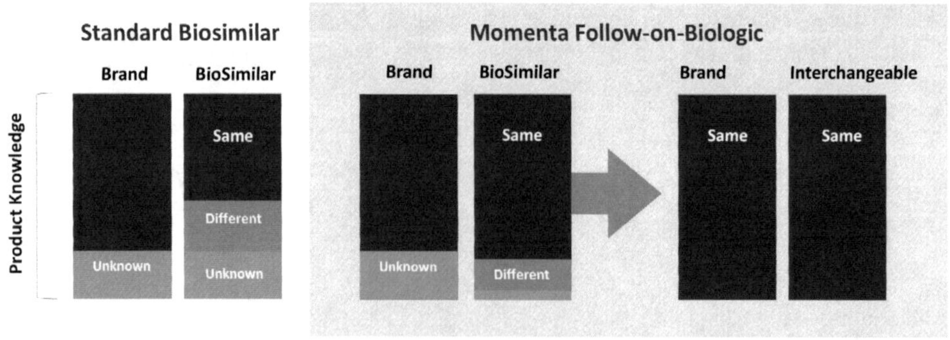


Remove **uncertainty.** Qualify **differences.** Demonstrate **equivalence.**

- **Thorough Product Characterization**
- **Manufacturing Process Design**
- **Product Control and Quality**

→

- **Increased POS for approval**
- **Reduced clinical requirements**
- **Opportunity for interchangeability**
- **Improved commercial differentiation**

## Considerations for Comparison

**Comparison should include the totality of evidence across multiple dimensions of the product**

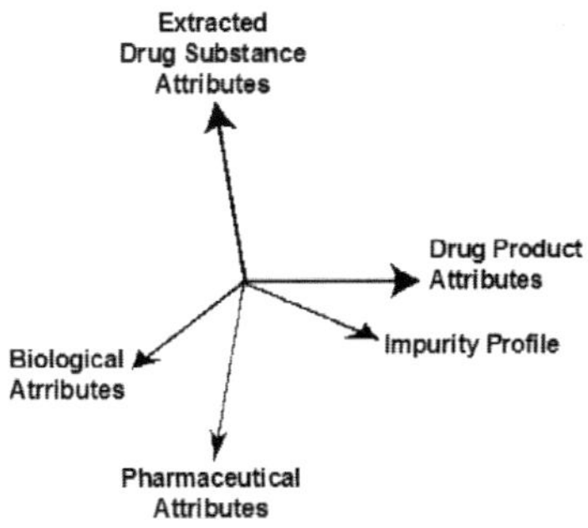


Comparison requires definition of:

- Which axis(es) / attributes to compare?
- How deep to look at that axis (resolution)?
- How to measure (quantify)?




## Considerations for Comparison

**Comparison should include the totality of evidence across multiple dimensions of the product**


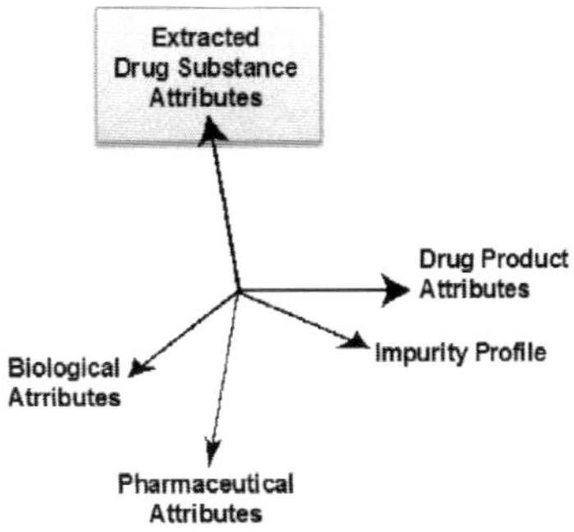

## Complex Biologics = complex mixture variants

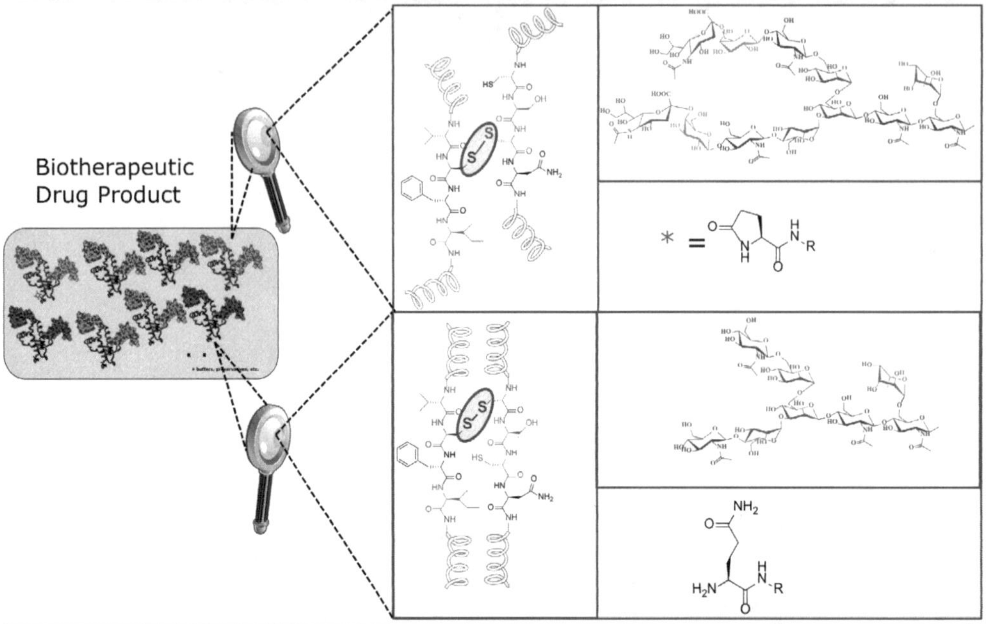




## Defining Structure: Challenges with Complex Mixtures

- Advanced analytical techniques offer limited "projections" of the complex mixture
- Structural attributes captured as ensemble averages
- Comparability based on "point to point"

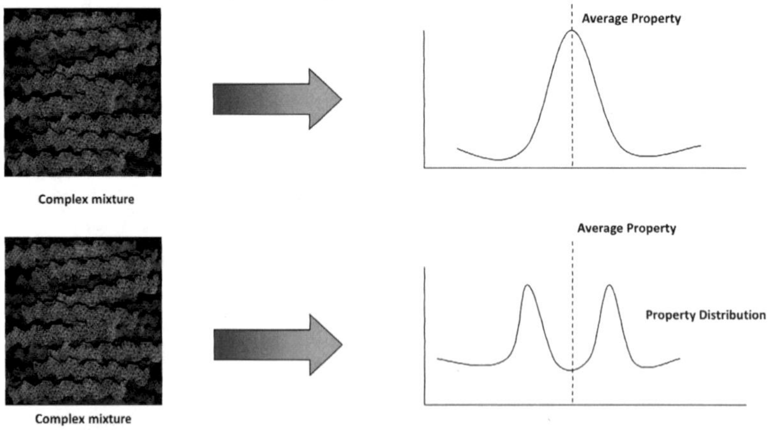

## Characterization of a complex product – Things to consider in Analytics

- Orthogonal views
  - Different techniques have blind spots and hence have to be augmented with orthogonal approaches
- Deep, high resolution and overlapping, going beyond "aggregate measures"
- Mass balance – capturing all of the components
- Sample prep challenges – introduction of Bias
- Not "point by point" collection of characteristics

⇓

**Key "fingerprints" or signatures that uniquely characterizes the reference product**

29 MOMENTA

MOMENTA

## Biologics Process Development

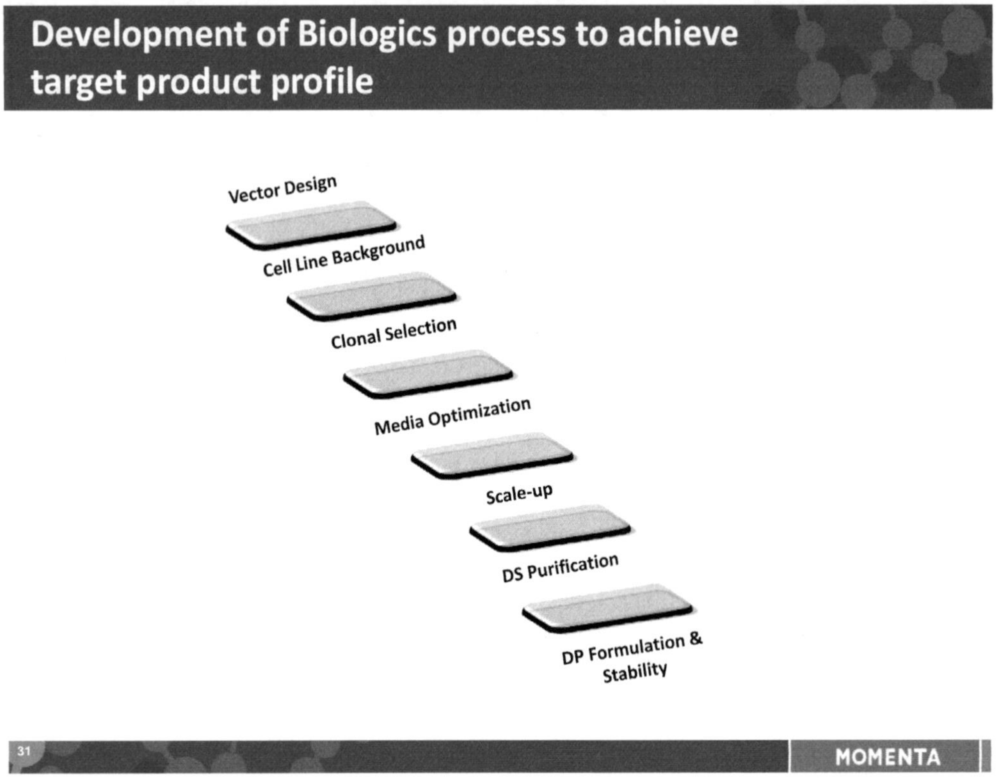
Development of Biologics process to achieve target product profile
Vector Design
Cell Line Background
Clonal Selection
Media Optimization
Scale-up
DS Purification
DP Formulation & Stability
31
MOMENTA

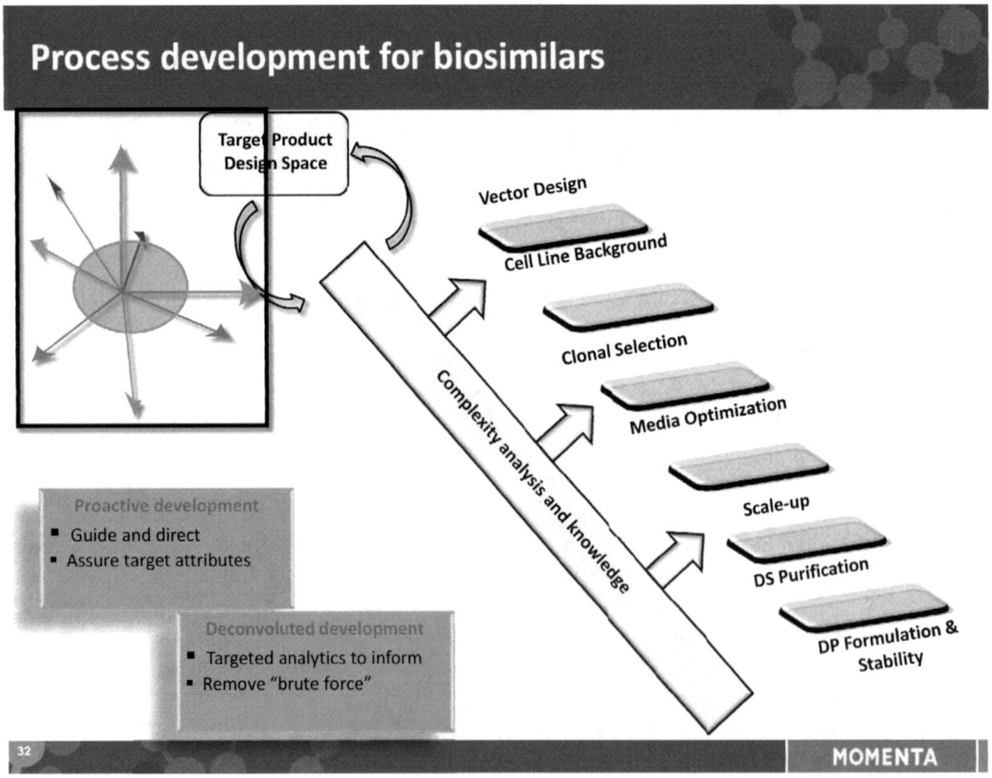
Process development for biosimilars
Target Product Design Space
Vector Design
Cell Line Background
Clonal Selection
Media Optimization
Scale-up
DS Purification
DP Formulation & Stability
Complexity analysis and knowledge
Proactive development
▪ Guide and direct
▪ Assure target attributes
Deconvoluted development
▪ Targeted analytics to inform
▪ Remove "brute force"
32
MOMENTA

## Targeted characterization delineates CHO cell line background

Genetic profiles

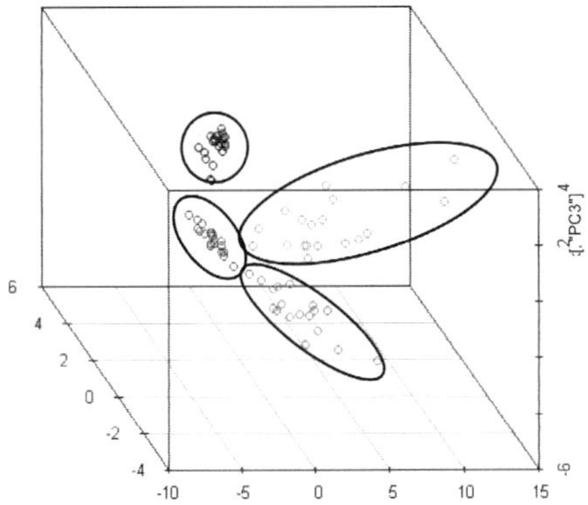


Product quality profiles

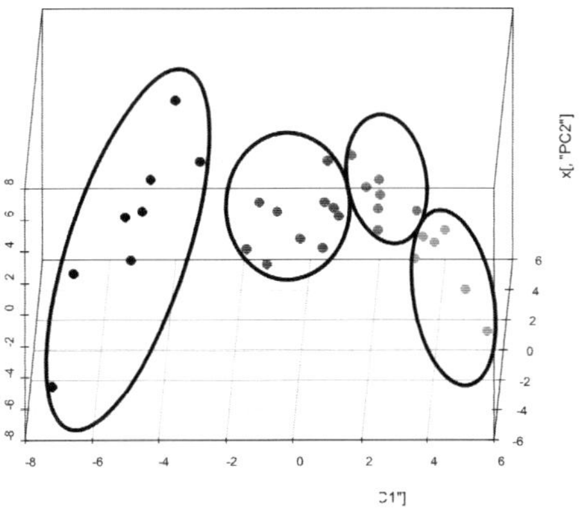


Cell line 1 - red
Cell line 2 - blue
Cell line 3 - black
Cell line 4 - green



## Example: Glycosylation - network map for understanding glycosylation (Cwire™)

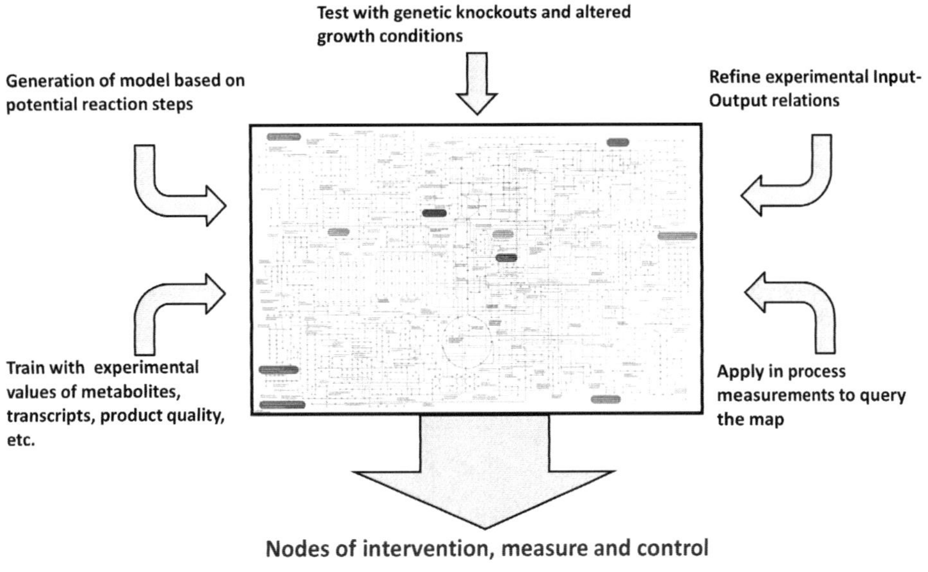

## Branch points that commit to specific "fingerprints"

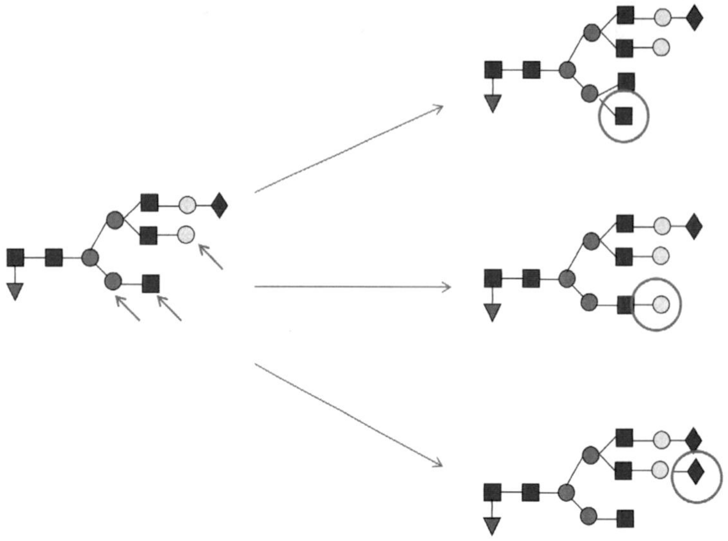

MOMENTA

## Momenta Technology: Analytics and Knowledge Applied Across Biologics Development

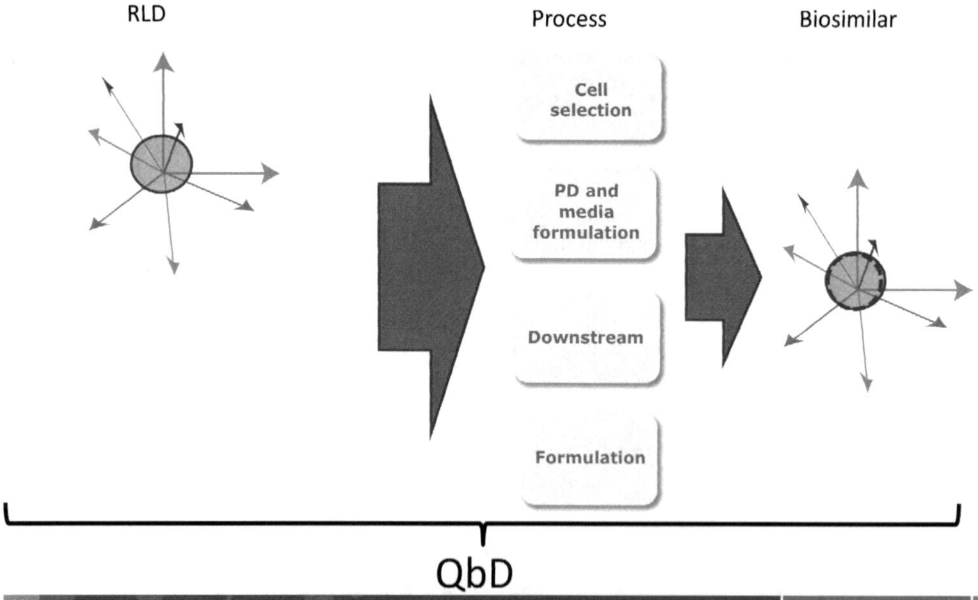


36 MOMENTA

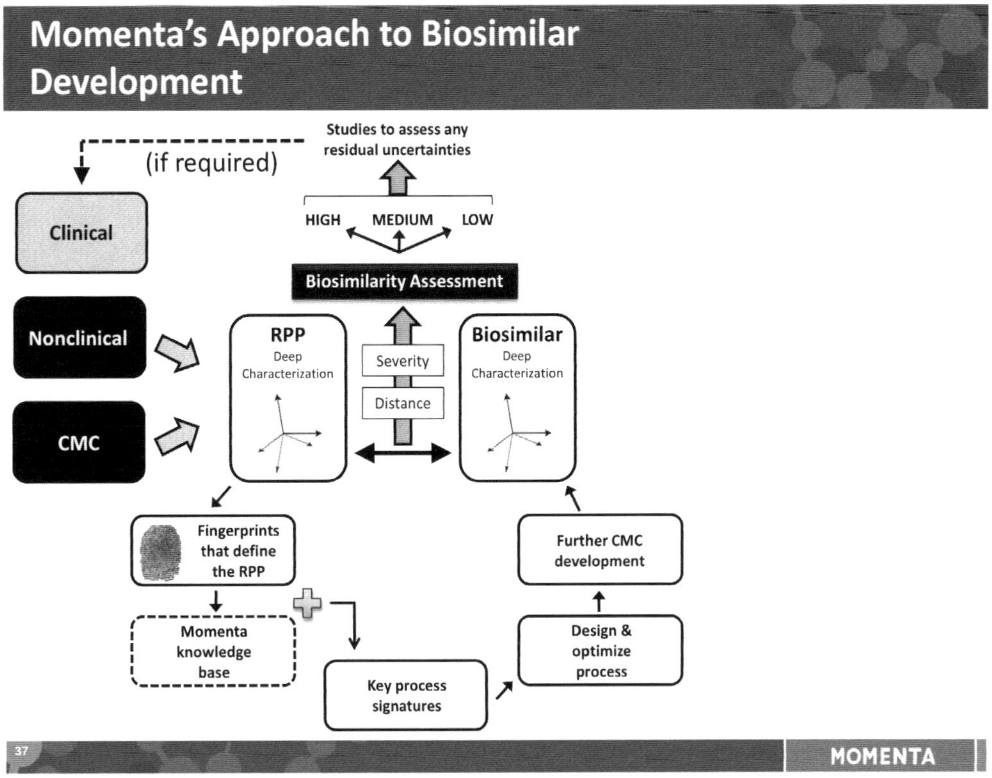
Momenta's Approach to Biosimilar Development
(if required)
Studies to assess any residual uncertainties
HIGH
MEDIUM
LOW
Clinical
Biosimilarity Assessment
Nonclinical
CMC
RPP
Deep Characterization
Severity
Distance
Biosimilar
Deep Characterization
Fingerprints that define the RPP
Momenta knowledge base
Key process signatures
Design & optimize process
Further CMC development
37
MOMENTA

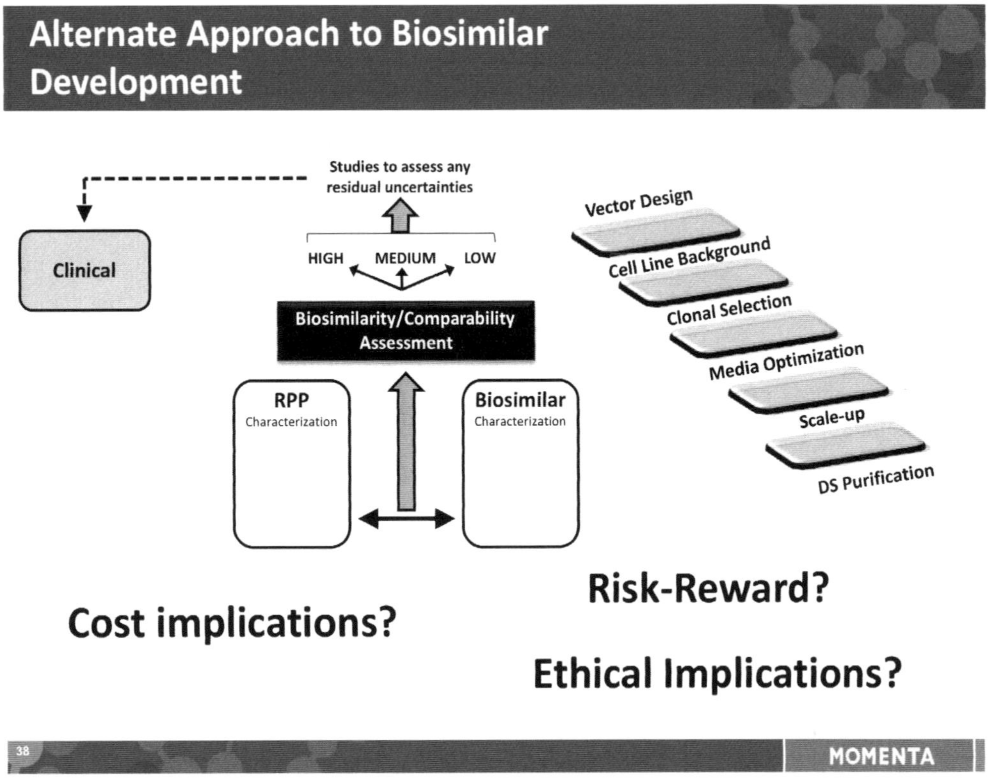
Alternate Approach to Biosimilar Development
Studies to assess any residual uncertainties
Clinical
HIGH
MEDIUM
LOW
Biosimilarity/Comparability Assessment
RPP
Characterization
Biosimilar
Characterization
Vector Design
Cell Line Background
Clonal Selection
Media Optimization
Scale-up
DS Purification
Cost implications?
Risk-Reward?
Ethical Implications?
38
MOMENTA

## Momenta's FOB Strategy: Not a Similar; a True Generic

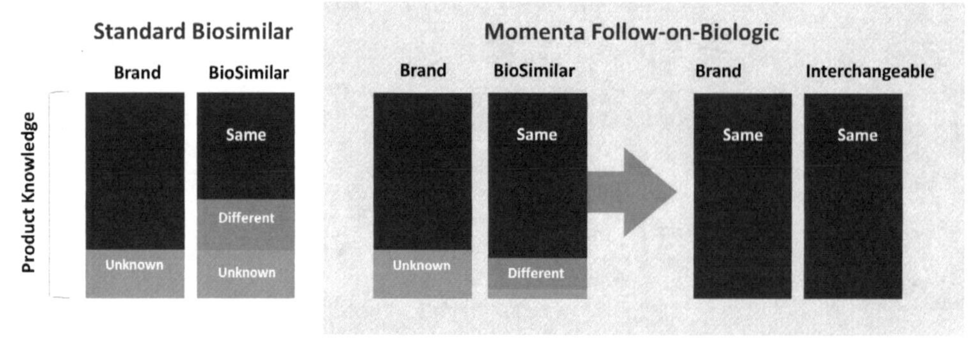


Remove **uncertainty.** Qualify **differences.** Demonstrate **equivalence.**

- Thorough Product Characterization
- Manufacturing Process Design
- Product Control and Quality

→

- Increased POS for approval
- Reduced clinical requirements
- Opportunity for interchangeability
- Improved commercial differentiation



## Scientific Principles Applied to the Review of Generic Enoxaparin are also Applicable to the Review of Biosimilar Products

The NEW ENGLAND JOURNAL of MEDICINE

Perspective

AUGUST 4, 2011

**Developing the Nation's Biosimilars Program**

Steven Kozlowski, M.D., Janet Woodcock, M.D., Karen Midthun, M.D., and Rachel Behrman Sherman, M.D., M.P.H.

*"...Such a "totality of the evidence" approach can also be applied to assessing biosimilars, **since it seems possible to exceed a current state-of-the-art analytic characterization by evaluating more attributes and combinations of attributes at greater sensitivities with multiple complementary methods.** There may be strategies that allow a "fingerprint"-like identification of very similar patterns in two different products. Such strategies were used in supporting the approval of a generic low-molecular-weight heparin product, enoxaparin **— which, though it differs from proteins in important ways, is structurally complex.** Although additional animal and clinical studies will generally be needed **for protein biosimilars for the foreseeable future,** the scope and extent of such studies may be reduced further if more extensive fingerprint-like characterization is used."*

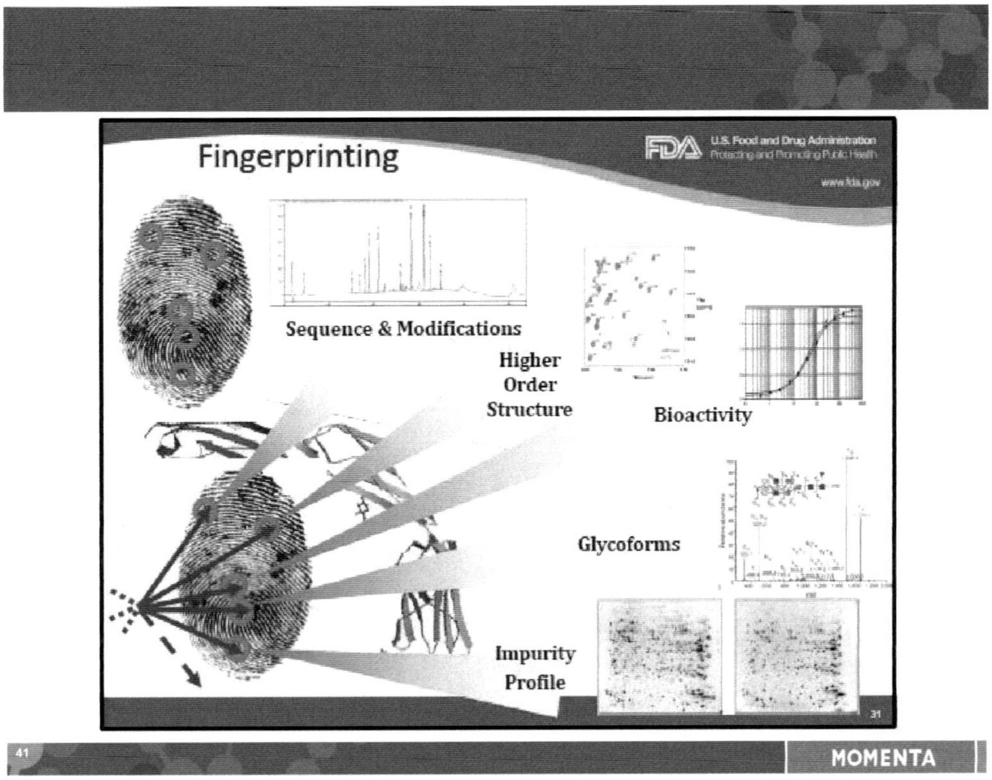
Fingerprinting
U.S. Food and Drug Administration
Protecting and Promoting Public Health
www.fda.gov
Sequence & Modifications
Higher Order Structure
Bioactivity
Glycoforms
Impurity Profile
31
41
MOMENTA

MOMENTA
Complexity in Biology

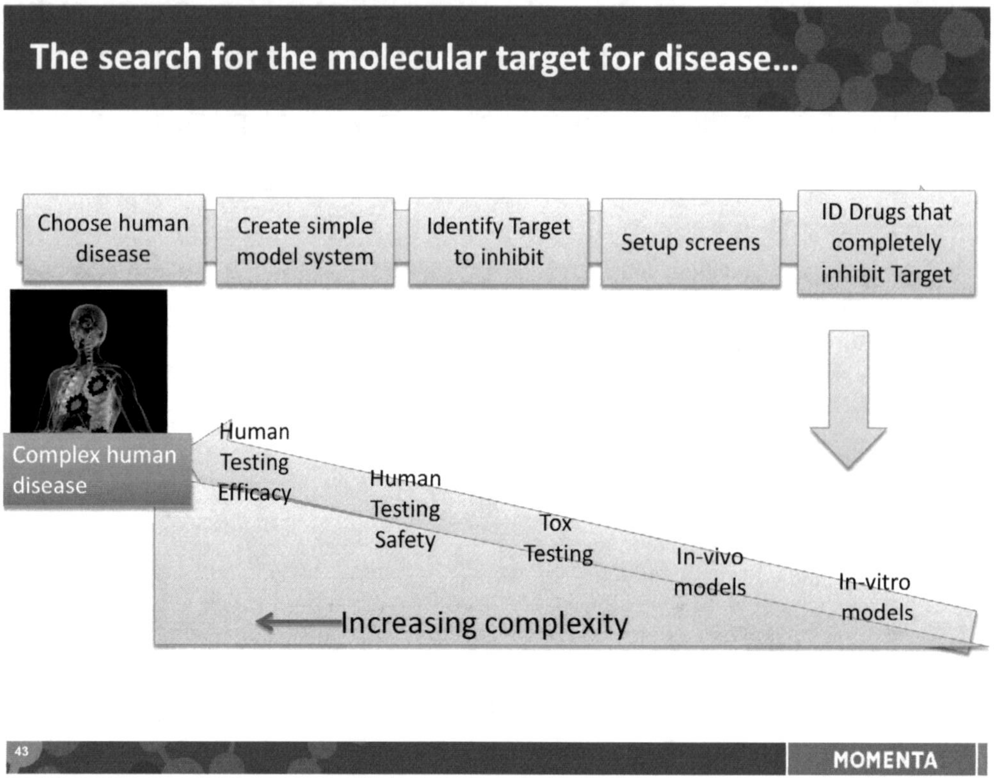
The search for the molecular target for disease...
Choose human disease
Create simple model system
Identify Target to inhibit
Setup screens
ID Drugs that completely inhibit Target
Complex human disease
Human Testing Efficacy
Human Testing Safety
Tox Testing
In-vivo models
In-vitro models
Increasing complexity
43
MOMENTA

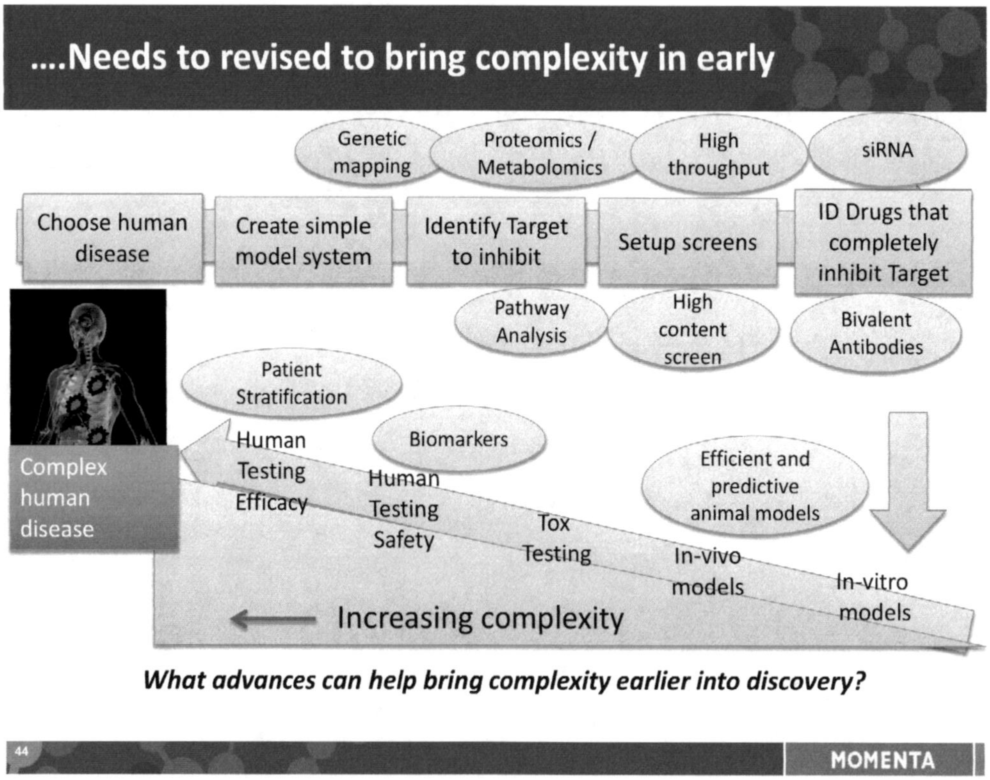
....Needs to revised to bring complexity in early
Genetic mapping
Proteomics / Metabolomics
High throughput
siRNA
Choose human disease
Create simple model system
Identify Target to inhibit
Setup screens
ID Drugs that completely inhibit Target
Pathway Analysis
High content screen
Bivalent Antibodies
Patient Stratification
Biomarkers
Efficient and predictive animal models
Complex human disease
Human Testing Efficacy
Human Testing Safety
Tox Testing
In-vivo models
In-vitro models
Increasing complexity
What advances can help bring complexity earlier into discovery?
44
MOMENTA

## In this era of "information age"....

B. Watterson, "*There's treasure everywhere*", Andrews and McMeel, 1996.



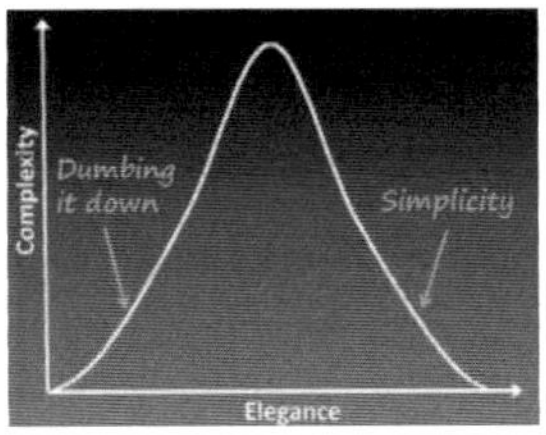


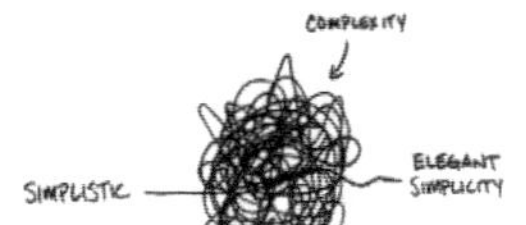


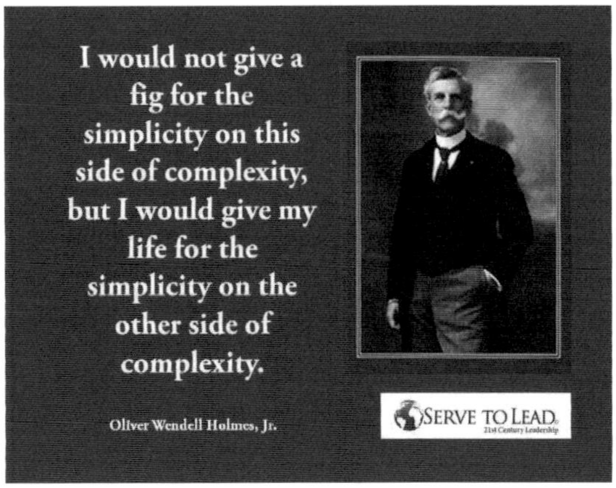

# Many Thanks…

**Contact:**
Ganesh V Kaundinya, PhD
CSO & VP Research, Co-Founder
Tel: +1 (617) 395-2844
E: ganesh@momentapharma.com

www.momentapharma.com

**Assessing the Criticality of Product Quality Attributes**

Flynn, G., Amgen

Originally presented on: 2/27/2013 8:05 - 8:35

Recombinant therapeutic proteins contain a variety of chemical and physical modifications, whose impact on safety or efficacy may be unknown. Many of the chemical conversions also occur in vivo, and knowledge of these reactions can be applied to assess the attribute's impact on the safety, potency and/or clearance of the drug. This talk will describe the development and use of tools designed to investigate attributes' biological impact, with emphasis on those tools that probe safety. A formal approach to ranking attributes' impact will also be presented.

CMC biologics
Right. On Time.

# QbD: A CMO's perspective on Gaining Knowledge Faster and Better

Clinton Weber, CMC Biologics



## Key Elements of QbD

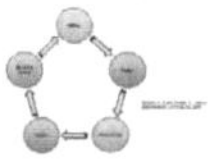

- Define
  - Define the Product and the Process
  - Key Product end goals: safety & Efficacy
  - Key Process Components: CMAs, CPPs & CQAs
- Design, Develop & Characterize
  - Prior knowledge / Literature / SMEs
    - Upstream Process
    - Downstream Process
    - Tools: DOEs, MVA, Models
- Verify/Validate
  - Process at the bench scale
  - Process at-scale
  - CQAs are consistently achieved
- Monitor & Control
  - Control strategy
  - CMAs and CPPs
  - PAT (Optional) & CPV
- Refine
  - Innovation
  - Continuous improvement of the process

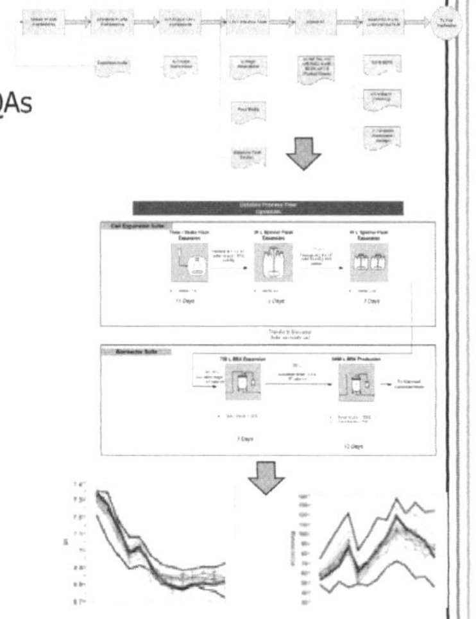

CMC biologics

## Steps of QbD by responsibility (12 Step Program) CMO vs. (Innovator)/Client

| Step # | Requirement | Description | Innovator vs. CMO |
|---|---|---|---|
| 1 | Defined TPP | Addresses patient needs | Client |
| 2 | Defined QTPP | Identified surrogate assays to address safety & efficacy of the product | Client |
| 3 | Establish Prior knowledge / experience | Related Products, Biologics Biosimilars & Biobetters | Both |
| | Knowledge gap from Risk Assessments | Related Products, Biologics Biosimilars & Biobetters | CMO |
| 4 | Defined CQAs | Critical Quality Attributes | Both |
| 5 | A manufacturing process that produces final product having the desired CQAs | Result: traditional 3 Conformance-run process | Client |
| | | Result: A QbD based process verification | CMO |
| 6 | Identified Critical Process Parameters | CPPs based on Risk Assessments | CMO |
| 8 | DOEs | Design of Experiments | CMO |
| 7 | Design Space | Based on CPPs and CQAs | CMO |
| 9 | Control Strategy | Process / Facility Controls | CMO |
| 10 | Risk Assessment | Application of Quality Risk Management principles | Both |
| 11 | Process Analytical Technology (optional) | In-line, on-line, at-line analysis of process steps | CMO |
| 12 | Continuous Process Verification | Monitor process parameters and critical quality attributes | CMO |

CMC biologics

## Case Study Differences

| | Traditional | QbD |
|---|---|---|
| Target Product Profile: | Not provided by Client | Provided by Client at Project Start |
| Quality Target Product Profile: | Not established until later | Developed early by both parties |
| Knowledge gap analysis: | Not formally performed | Formally performed as part of QRM |
| Critical Quality Attributes: | Not identified until later | Mutually developed and agreed upon |
| Critical Process Parameters: | Incorrectly identified (not robust) | Based on prior process knowledge and experience, identified by Risk Assessment |
| Use of Design Space: | DOE studies randomly performed – no Design Space established | DOE studies performed – Design Space established for most unit operations |
| Process Verification at PD stage: | Scale Down model established late | Scale Down model established early |
| Process Verification at-Scale: | Three conformance runs | Considering 2 Confirmatory Runs |
| Process Control Strategy: | Through input & output parameters | Through input & output parameters |
| QbD based regulatory submission: | Not done | Developed by both parties |
| Continued Process Verification: | Initiated, but very large | Expected to be optimal size |
| Post verification improvements: | Not done because of regulatory constraints | Planning on several |

CMC biologics

# Case Study – QbD vs. Characterization

BACKGROUND

- Two different products, two different clients
- Both products are batch production cell culture processes
- Both products came with RCB only (clone selection performed by CMO)
- Both products are complex proteins (not Mab)
- **Contractual differences determined course for QbD**
- **Total Knowledge = identification of CQA's/CPP's with ranges justified (same for both products)**
- QbD Case Study is ongoing

# Simple Illustration of the QbD Goal

Traditional Approach

Design Space Approach

Courtesy of Alex Brindle, NNE

# Examples of Tools to drive DOE design

P-Diagram (inputs vs. outputs)

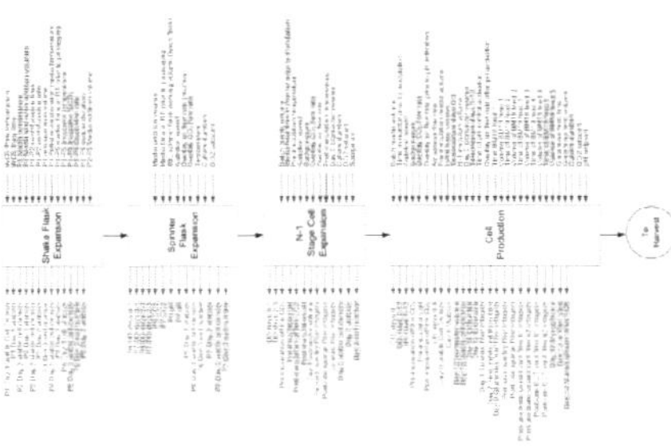

Risk Assessment Mitigation Matrix (RAMM)

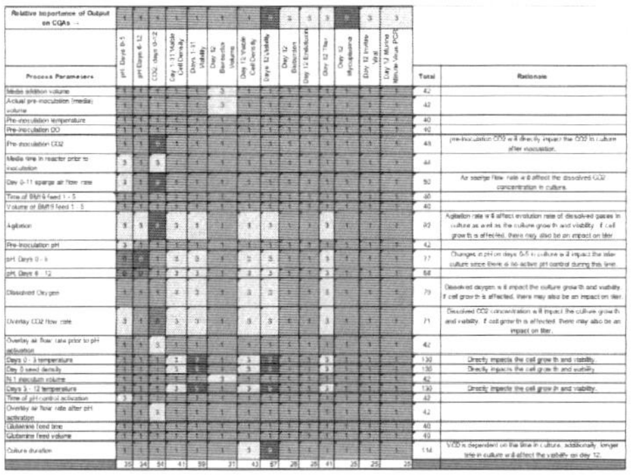

7

CMC biologics

# Case Study – QbD vs. Characterization

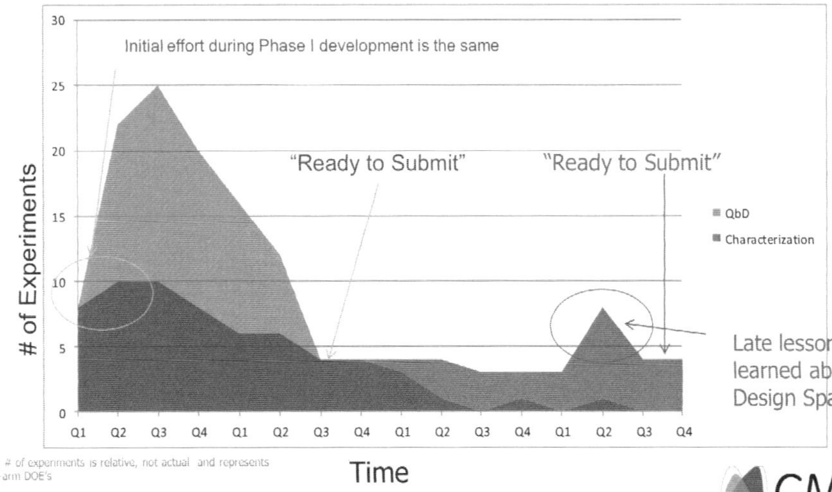


Note: # of experiments is relative, not actual and represents multi-arm DOE's

CMC biologics

# Case Study – Improving DOE's

Traditional Characterization = 3 conditions with 2 factors (8 flasks)

Process Optimization Experimental Design

| Flask | Seed Density ($x10^6$) | Temp Shift (Day) | Temp Shift Set-point (°C) |
|---|---|---|---|
| 1 | 0.50 | 3 | 30 |
| 2 | 0.50 | 3 | 34 |
| 3 | 0.50 | 5 | 32 |
| 4 | 0.50 | 7 | 30 |
| 5 | 0.50 | 7 | 34 |
| 6 | 0.75 | 3 | 32 |
| 7 | 0.75 | 5 | 30 |
| 8 | 0.75 | 5 | 32 |
| 9 | 0.75 | 5 | 32 |
| 10 | 0.75 | 5 | 34 |
| 11 | 0.75 | 7 | 32 |
| 12 | 1.00 | 3 | 30 |
| 13 | 1.00 | 3 | 34 |
| 14 | 1.00 | 5 | 32 |
| 15 | 1.00 | 7 | 30 |
| 16 | 1.00 | 7 | 34 |

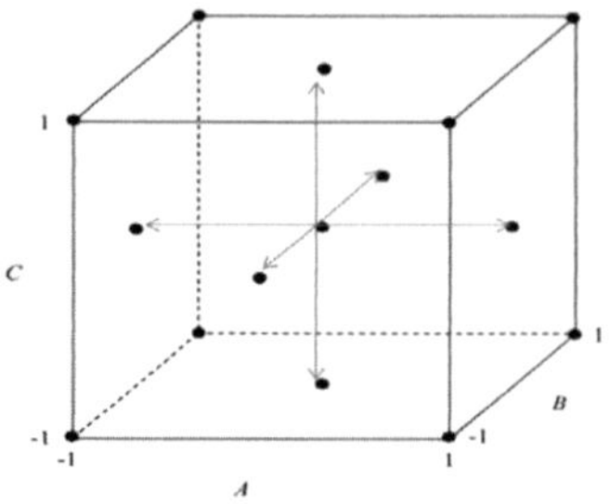


Central Composite Design with Center Point Replicates

CMC biologics

# Results (JMP Analysis)

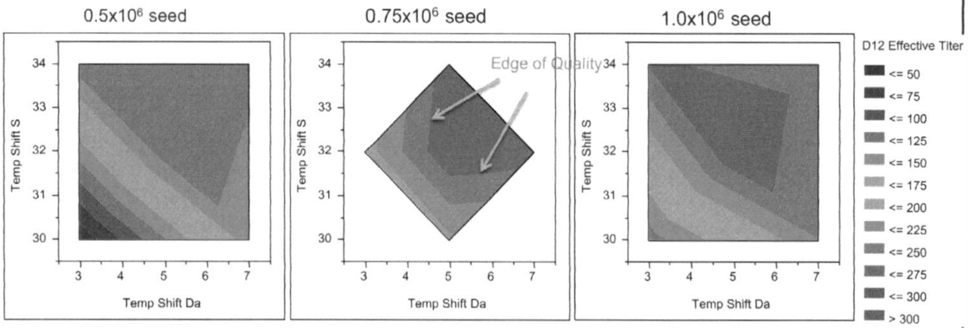


- Statistically significant factors that effect D12 effective titer
  - Temp shift time ($p = 0.0001$)
  - Temp shift set point ($p < 0.0001$)
  - Seed density ($p = 0.0264$)
  - Temp shift day interaction with temp shift set point ($p < 0.0001$)
  - Temp shift day quadratic effect ($p = 0.0047$)
  - Temp shift set point quadratic ($p = 0.0067$)
- Experimental power:
  - 88.46% chance of detecting a change of 15 mg/L

| Term | Orthog Estimate |
|---|---|
| Temp Shift SP(30,34) | 35.53550 |
| Temp Shift Day*Temp Shift SP | -28.66938 |
| Temp Shift Day(3,7) | 27.03644 |
| Temp Shift Day*Temp Shift Day | -25.56630 |
| Temp Shift SP*Temp Shift SP | -14.78692 |
| Seed Density(0.5,1) | 11.18882 |

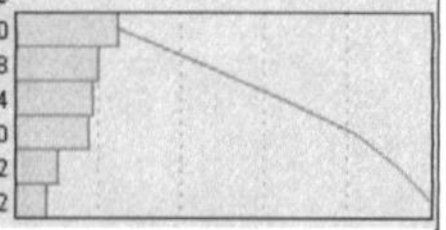

10

# Case Study – Optimizing Ranges vs. "Range Finding"

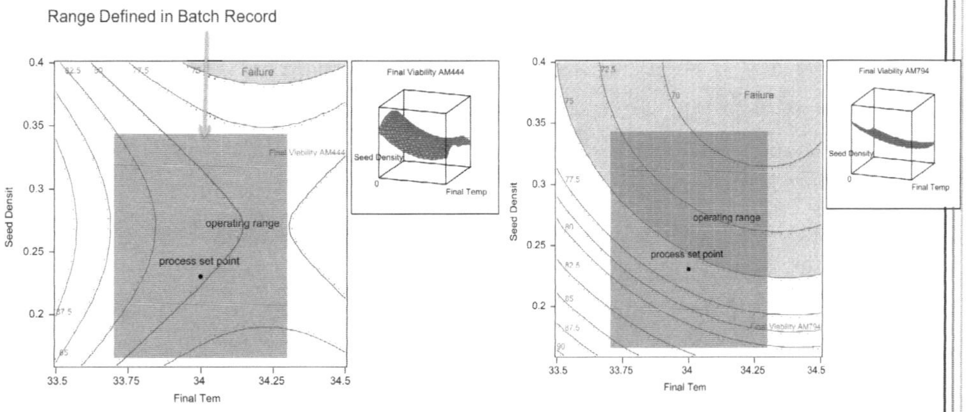


Raw Material Lot influence

CMC biologics

# Case Study – QbD vs. Characterization

MOST IMPORTANT LESSONS LEARNED

- Smart DOE's are difficult, but good Risk Assessment tools and initial parameter screening studies can improve efficiency of the DOE's
- Early/Starting with QTPP is vital to a QbD approach
- Late Delivery of CQA's is very costly – delays Method Qualification/Validation and causes late process design changes
- A good QbD program will shift (non-clinical) critical path from process development to Method development
- Raw Material contribution to variation should not be understimated
- The characterization approach results in a much larger CPV Program/Protocol – increasing ongoing costs from sampling/analysis
- Time is money for CMO's, savings come from increasing DOE size

CMC biologics

# Case Study – QbD vs. Characterization

SUMMARY

- Having well-understood CQA's is key to success
- Total cost of development (# experiments) is within 30% of each other
- QbD approach resulted in 70% less deviations during first GMP Manufacturing Runs
- Up to 18 months savings in timeline (not in cost factor, but should be)
- End result = more knowledge, faster and with improved robustness

CMC biologics

**Control Strategies for Antibody-Drug Conjugates Drug Product Manufacturing**

Jiang , S., Seattle Genetics

Originally presented on: 2/27/2013 9:05 - 9:35

Antibody-drug conjugates (ADCs) have unique attributes that pose challenges in the development and manufacturing. A control strategy for ADC drug product (DP)-related critical quality attributes (CQA) was developed based on the assessment of product quality attribute criticality and process capability. Appropriate and multiple levels of control assure these quality attributes can be consistently maintained by the DP manufacturing process..

# MedImmune's Experience in the FDA QbD Pilot Program: The Story So Far...

P. David Robbins, Guillermo Miró-Quesada, & Melia J. Grim

AccBio/QbD 2013

SBE's 3rd International Conference on Accelerating Biopharmaceutical Development Including Quality by Design in Biologics (QbD)

QbD Roundtable

February 27, 2013

## Why should MedImmune embrace Quality by Design?

- Structured risk-based approach to development
  - Prioritization of activities to address areas of greatest risk
  - Increased efficiency to help us deliver MedImmune's growing late-stage pipeline
- Improved process understanding resulting in more robust processes, more effective control strategies, and greater assurance of quality
  - Reduced technical risks for projects, with fewer late stage surprises
  - Facilitates investigation of manufacturing deviations
  - Solid framework for continuous improvement
- Systematic approach helps the regulators better understand company's strategy, enhances scientific foundation for review
- Potential manufacturing flexibility through Design Space
- Consistent with best practices at peer companies and with current FDA expectations

MedImmune

MedImmune's Experience in the QbD Pilot Program

## MedImmune's Approach to Developing QbD

- Develop strategies and systems that are practical and sustainable
  - Flexible approaches (*e.g.*, risk assessment tools) applicable to all projects (not two systems QbD vs. non-QbD)
  - Case-by-case determination of which QbD elements can benefit a particular filing
    - Do we claim a Design Space? What do we expect to get out of it?
- Build upon existing good practices and scientific knowledge
  - More systematic approaches improve consistency and drive efficiency
  - Systems designed to focus and clarify thinking, not cookie-cutter approaches that stifle thinking
- Primary objectives:
  - Better process/product understanding
  - Drive efficiency through risk-based prioritization
    - Manufacturing flexibility also of interest but not required to justify value of QbD
- "Think like a regulator"
  - How would you view the proposal if you were the FDA reviewer?

MedImmune

MedImmune's Experience in the QbD Pilot Program

## FDA QbD Pilot Program

- Designed by FDA to facilitate the implementation of QbD principles
  - Small molecule program (2005)
    - MedImmune's parent AstraZeneca took part
  - Biologics program launched by Office of Biotechnology Products in 2008
  - MedImmune application accepted January 2011
- Participants are granted additional meetings to present their QbD approaches **applied to a real project**
  - De-risk inclusion of QbD elements in final BLA submissions
  - Not possible in industry consortiums case studies
    - A-Mab Case Study, an FDA-Industry Consortium in which MedImmune participated (2008)
  - Information shared in Pilot Program meetings is non-binding
- 1st meeting July 2012
  - Focus on Quality Attribute Risk Assessment and Severity Assessments
  - Outcome: FDA generally supportive of MedImmune proposals and approach
- 4 Subsequent Meetings Agreed Upon by FDA Once/Year

MedImmune

MedImmune's Experience in the QbD Pilot Program

## Severity Assessment and Development of Control Strategy at Various Stages of Product Development

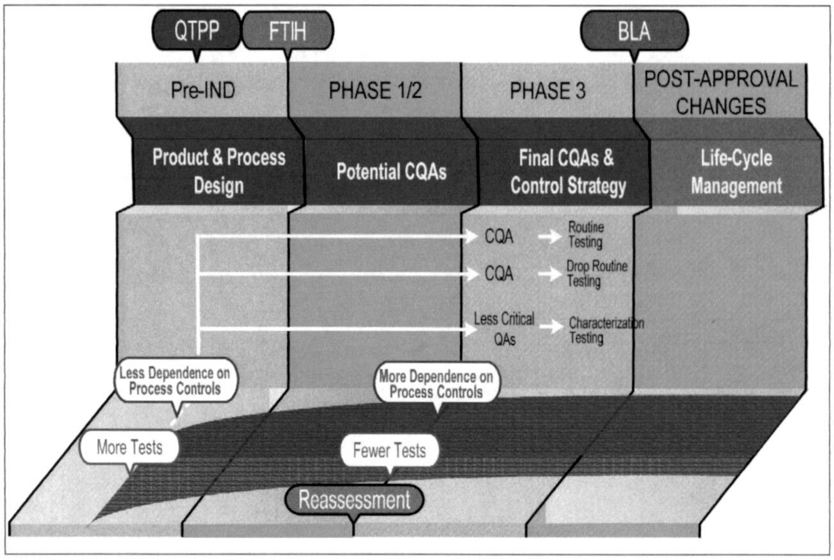


MedImmune

MedImmune's Experience in the QbD Pilot Program

## High-Level Strategy for Quality Attribute Risk Assessment

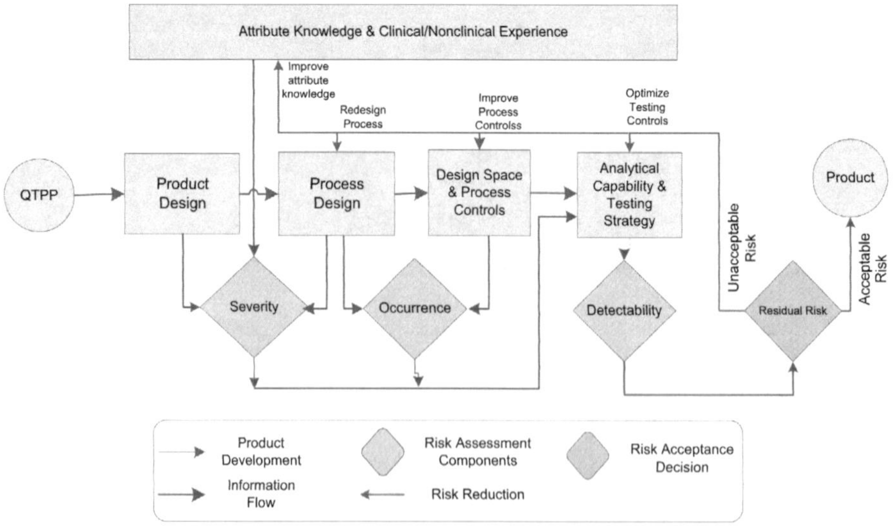


QTPP = Quality Target Product Profile

MedImmune

MedImmune's Experience in the QbD Pilot Program

## Quality Attribute Risk Assessment via FMEA

- Assess each Quality Attribute in the three categories below
  - Risk assessment approach (FMEA) developed as refinement of earlier proposals in A-Mab QbD case study, and informed by approaches presented by other companies
  - The combination of the three categories represents the residual risk to the patient arising from each quality attribute

| Definitions of Severity, Occurrence and Detectability | | |
|---|---|---|
| | **IEC-60812 Definition** | **Definition Adapted for Quality Attribute Risk Assessment** |
| Severity | "...an estimate of how strongly the effects of the failure will affect the system or the user..." | Impact on patient's safety and efficacy when dosed with product with quality attribute outside of its appropriate limit, range or distribution |
| Occurrence | "...probability of occurrence of a failure mode..." | Likelihood that a quality attribute will be outside of its appropriate limit, range or distribution |
| Detectability (Detection) | "...an estimate of the chance to identify and eliminate the failure before the system or customer is affected..." | A measure of the ability to identify whether a quality attribute is outside of its appropriate limit, range or distribution prior to patient dosing |

MedImmune

MedImmune's Experience in the QbD Pilot Program

## Risk-based control strategy development

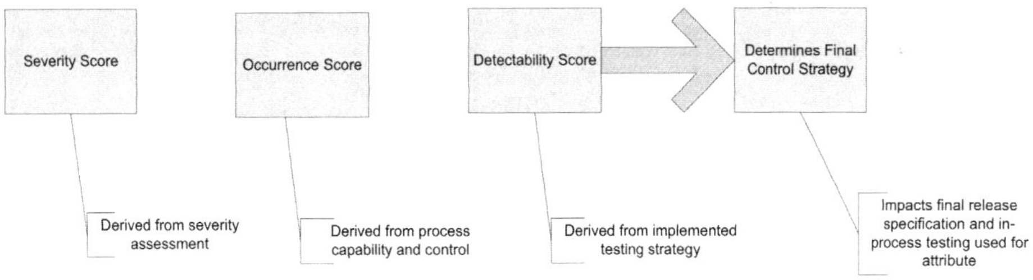


*Objective is to show that manufacturing process and control strategy reduce risk to acceptable levels for all quality attributes*

*Provides a systematic, scientifically based approach to control of quality attributes*

MedImmune

MedImmune's Experience in the QbD Pilot Program

## RPN = S×O×D as measure of residual risk:

preliminary assessment (S), process capability and control strategy not taken into account

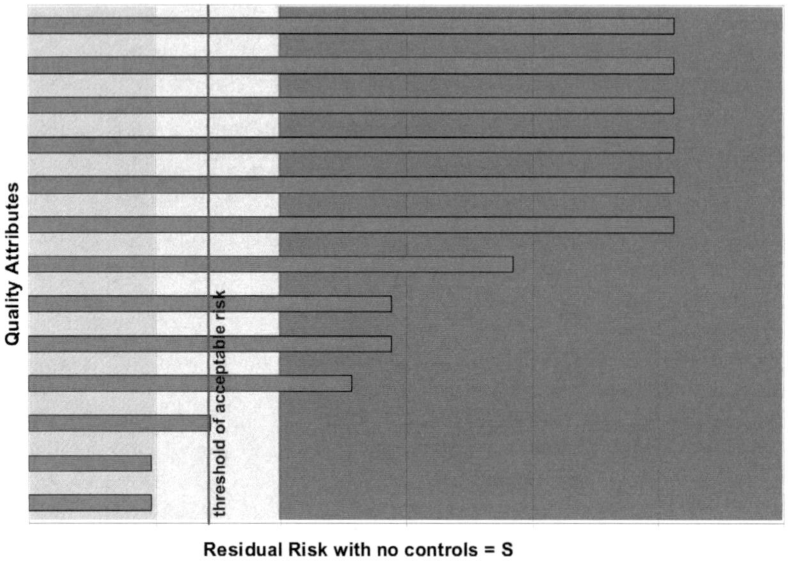


MedImmune

MedImmune's Experience in the QbD Pilot Program

## RPN = S×O×D as measure of residual risk:

Design Space and Process Controls (Occurrence); no testing

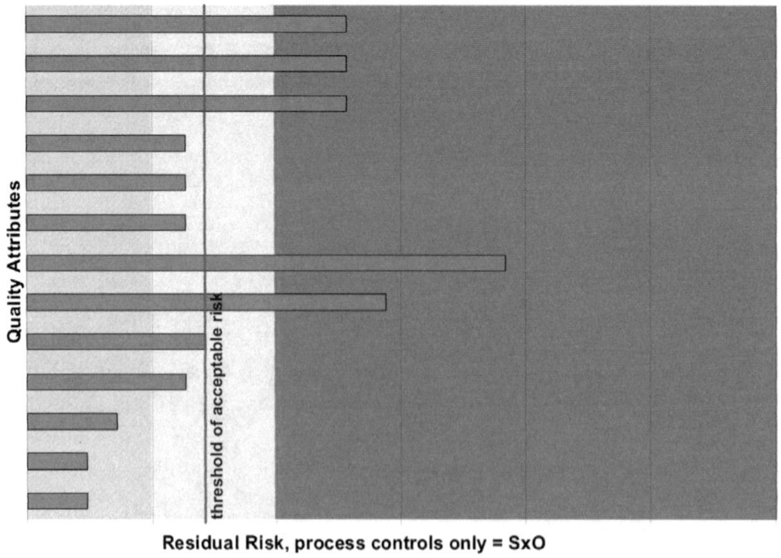


MedImmune

MedImmune's Experience in the QbD Pilot Program

# RPN = S×O×D as measure of residual risk:
## full Control Strategy, Process (O) and Testing (D) Controls

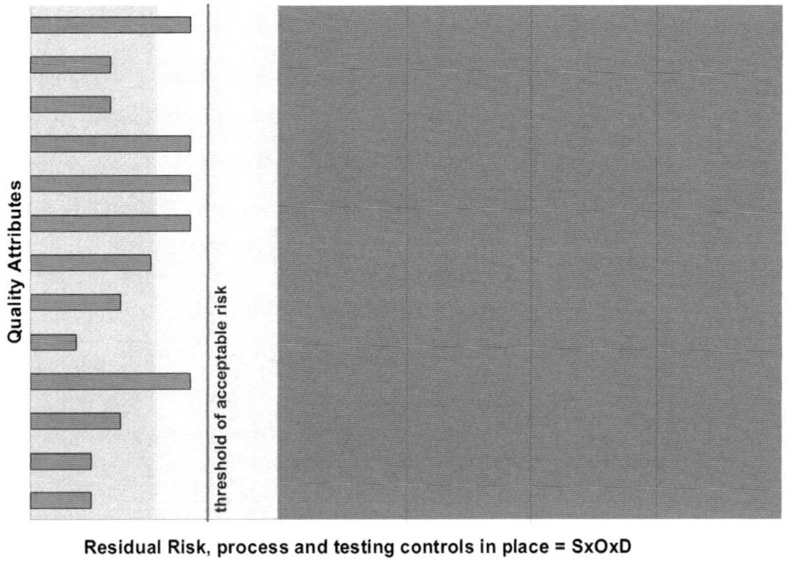




# *Severity* Assessments
# (primary focus of 1st FDA meeting in 2012)

- Measure of **a specific attribute's** potential to impact the safety and efficacy of the product
  - Four impact categories:
    - Bioactivity
    - PK/PD
    - Immunogenicity
    - Safety (exc. Immunogenicity)
  - Does not take into account process capability
- The severity score is determined based on the potential ***impact*** and the ***uncertainty*** in the knowledge regarding the impact.
  - Uncertainty depends mainly on the source of knowledge
    - Literature > *in vitro* studies > non-clinical > clinical data
    - Similar products > Product specific
  - Uncertainty elevates the severity score to allow for a more severe impact than the one observed

# Impact and Uncertainty

## Impact Levels (potential to cause harm to patient)

| Level | Definition |
|---|---|
| Very high | Life-threatening or irreversible impact (irreversible disease progression, other irreversible effects) |
| High | Non-life threatening and reversible impact (reversible disease progression, other reversible effects) |
| Medium | Tolerable, manageable and transient impact |
| Low | Small but detectable impact |
| None | Negligible or no impact |

## Uncertainty of Impact assessment

| Level | Definition |
|---|---|
| High | Impact of specific attribute is based on relevant scientific literature |
| Medium | Impact of specific attribute predicted from laboratory or nonclinical studies with this molecule, or data from laboratory, nonclinical or clinical studies with related molecule in a relevant model/patient population |
| Low | Impact of specific attribute established from clinical studies with this molecule (or process-related impurities, or contaminants) |

MedImmune

MedImmune's Experience in the QbD Pilot Program

# Impact scoring for Each Impact Category

| | Impact Categories | | | |
|---|---|---|---|---|
| **Level of Potential Patient Harm** | **Bioactivity** | **PK/PD** | **Safety** | **Immunogenicity** |
| Very High | Impact on bioactivity with potential life-threatening effects or irreversible disease progression[a] | Impact on PK/PD with potential life-threatening effects or irreversible disease progression due to loss of efficacy[a] | Irreversible or life-threatening AE | Immunogenic response observed with potential life-threatening effects or irreversible disease progression |
| High | Impact on bioactivity with non-life-threatening loss of efficacy and reversible or no disease progression | Change in PK/PD with non-life-threatening loss of efficacy and reversible or no disease progression | Reversible and non-life-threatening AE, including those that result in stopping treatment or requiring hospitalization | Immunogenic response observed with non-life-threatening effects and confers limits on safety or efficacy |
| Medium | Tolerable impact on bioactivity within the limits required for efficacy | Tolerable impact on PK/PD within the limits required for efficacy | Manageable and transient AE that does not require hospitalization; minimal intervention required | Immunogenic response observed with tolerable effects on safety or efficacy |
| Low | Small but detectable change in bioactivity | Small but detectable change in PK/PD | Minor and transient AE; asymptomatic or mild symptoms; intervention not required | Immunogenic response observed with minimal in vivo effects |
| None | Non-detectable impact on bioactivity | Non-detectable impact on PK/PD | No AE detected | Immunogenic response not observed |

AE: Adverse event; PD: Pharmacodynamics; PK: Pharmacokinetics

[a] Loss of efficacy can result in life-threatening effects or irreversible disease progression depending on indication

MedImmune

MedImmune's Experience in the QbD Pilot Program

## Severity Scoring and Critical Quality Attributes (CQAs)

| | | Uncertainty | | |
|---|---|---|---|---|
| | | **Low** | **Medium** | **High** |
| **Impact** [a] | Very High | 32 | 32 | 32 |
| | High | 16 | 24 | 32 |
| | Medium | 8 | 12 | 18 |
| | Low | 4 | 6 | 9 |
| | None | 1 | 2 | 3 |

Red/bold= CQA; Green= less-critical quality attribute

**Critical Quality Attribute (ICH Q8R2):** A physical, chemical, biological or microbiological property or characteristic **that should be within an appropriate limit, range, or distribution** to ensure the **desired product quality**.

MedImmune

MedImmune's Experience in the QbD Pilot Program

## Overall Summary from July 2012 FDA Meeting

- Topics covered:
  - General Overview of Program (including agreement on meeting timings/topics)
  - Overview of the QbD Quality Attribute Risk Assessment
  - Severity Assessment
- General acceptance with the questions we asked
  - Not surprising, conservative approach to Severity assessment and CQAs was built on previous FDA feedback from industry case studies, and FDA/industry experience in Pilot Program
- Feedback on amount of information presented:
  - Acceptable for the 1st meeting Briefing Document, but would expect to see significantly more detail in a BLA

MedImmune

MedImmune's Experience in the QbD Pilot Program

## FDA feedback:

◆ **On Severity scoring:**

- *Use of impact and uncertainty analysis to derive Severity score is appropriate*
- *Conservative approach (using the highest impact score among various categories for each quality attribute) was well received*
- *More detail will be needed in BLA on method for determining impact or uncertainty levels*
- *Threshold of ≥ 9 for CQA acceptable*

◆ **On risk-based control strategy:**

- *FDA agreed that the risk of a CQA with a high severity score can be mitigated by process capability to such an extent that continued routine testing may not be required.*
- *Specific assessments regarding the removal of routine testing would be a BLA review issue*

MedImmune

MedImmune's Experience in the QbD Pilot Program

## Additional FDA feedback:

◆ Interactions between quality attributes may need to be studied further

◆ It might be necessary to assess site specific modifications, e.g., deamidation, oxidation, and glycation, rather than a general assessment of these modifications to the whole antibody

◆ Exclusion of attributes from Severity assessment must be justified in the BLA.

◆ If *in vitro* data are used, the link to *in vivo* clinical outcomes should be justified.

◆ FDA prefers quantitative impact levels or more tangible definitions of terms such as "small" or "tolerable" change.

◆ Use of prior knowledge from other mAbs is acceptable, but must be justified. Considerations should include:

- cell binding mechanisms
- different patient populations

◆ Route of administration should be taken into account in the risk assessments.

MedImmune

MedImmune's Experience in the QbD Pilot Program

# Quality by Design at Genentech & Roche in 2013

## *Lessons Learned from Two Case Studies in the FDA QbD Biotech Pilot*

*Lynne Krummen, Ph.D.*
*Global Head Roche Technical Regulatory, Biologics*
*Global Lead, Roche QbD Core Team*

## Case Study Summary

- Genentech & Roche entered 2 submissions in the FDA QbD Biotech Pilot in 2009
  - eCP (expanded comparability protocol) for multi-product, multi-site Drug Substance transfers
  - Original BLA for pertuzumab (Perjeta)
- eCP was US specific, approved in 2010
- Perjeta was filled globally – several global HA received pre-submission QbD overviews
  - All filings contained the same process and parameter descriptions and Control Strategy proposals.
  - Filings in some ICH regions also contained a proposal for Design Space
  - US and FDA conducted a collaborative review. PMDA participated as an observer
- Perjeta has been approved in US and several other countries, EU and global approvals are pending
  - QbD–based proposals related to control system design and process parameter ranges were approved
  - US and EU did not accept Design Space

## Focus of Today's Talk

- Lessons learned from development and implementation of GNE/Roche large molecule QbD program & participation in the FDA Pilot Program
- Current perspectives on benefits of QbD implementation
- Thoughts regarding perspectives on risk-tolerance and "regulatory flexibility"

## Perjeta and Expanded Change Protocol QbD Pilots

- Key products of the QbD pilot projects were risk-assessment tools and practices
- Intended to analyze and classify risks commonly associated with process development, control strategy and post-approval site-transfers
- Provide consistent basis for Subject Matter Expert positions and justification of use of platform and literature knowledge
- Create a common language and help clarify the decision making framework
- Provide detailed justification for associated scoring and decision criteria
- RA tools work together and create a complex analysis that fully captured SME thinking

***RA methodology was the main focus of pilot meetings***

## Case Study #1: Perjeta BLA
### Key Questions in Design of the BLA Control System Drove Development of Several Risk Assessment Tools & a "PALM" plan

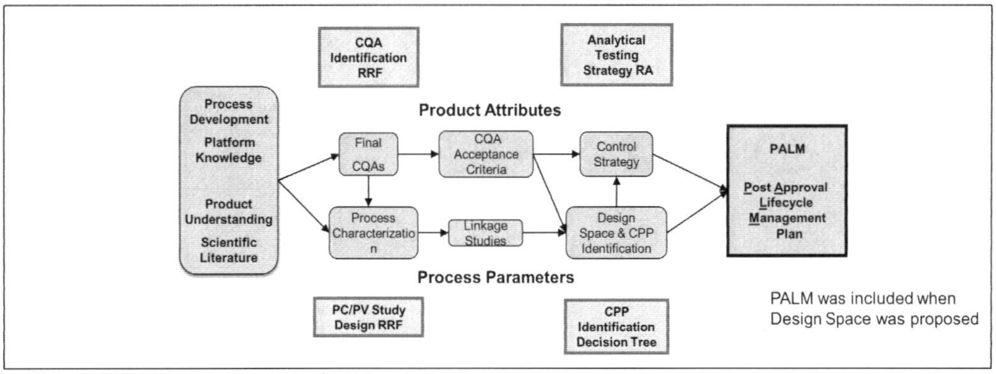


- What attributes are important?
- What are the appropriate limits for each attribute? Should they be based on patient or process – or both?
- Does the process provide attribute control? Is the attribute stable?
- What should be tested?
- How will control of the attribute and process be ensured in the future?

## Key Concerns Encountered

- How much residual risk is too much?
  - To some extent, risks are cumulative
- Could there be unforeseen consequences?
- How can the Agency be sure that post-approval risk management programs will be effective?
- High level of scrutiny to any decision that potentially shifts oversight from "pre-approval required" to "managed in the QMS" and reviewed upon notification or inspection
  - Bar for assessment of non-criticality
  - Residual uncertainty in Scale Down Model
  - Attributes Tested on Lot Release –vs- during Process Monitoring (esp. where monitoring is less frequent)

## Multiple Conservative Assumptions and Practices Were Incorporated to Address Uncertainty & Residual Risk

Roche

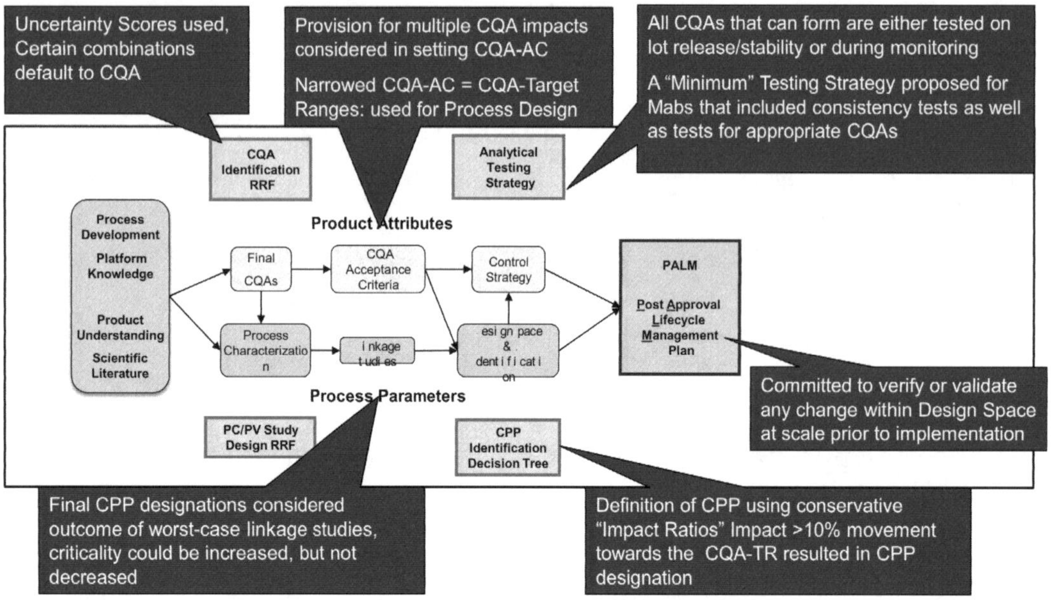


**Items were viewed as "Helpful"**

## QbD Objectives & Outcomes for Perjeta BLA

Roche

### There were 3 main objectives for the Perjeta filing:

1) Reduce redundant, non-value added QC testing based on risk assessments

2) Widen acceptance criteria for some CQAs based on product and platform understanding of patient impact

3) Obtain approval of a Design Space to facilitate management of post-approval changes without FDA pre-approval

### Accomplishments

**Reduced Control System Testing** ✓

- Accepted justifications to remove redundant or low/no value tests
- Created a category of "Comparability & Monitoring" (CaM) testing for moderate CQAs with high process capability

**Wider CQA-Acceptance Criteria** ✓

- Accepted justifications for proposed CQA-AC that extended well beyond clinical experience in some cases

## Perjeta QbD "Misses"

- **Sponsor's argument that ADCC was not a MoA was not accepted**
  - Not all CQAs were identified
    - Not all CPPs that impacted those CQAs were identified
      - Control system proposals related to those CQAs were not appropriate

- **Sponsor's Design Space Proposal was not accepted**

  Missing CPPs related to ADCC MoA were a factor, but not the only reason:

  **Design Space definition in BLA: "combination of all CPPs" was unclear**

  (**Risk**: unit operations w/o CPPs get lost, lack of oversight of non-CPPs)

  **Remaining lack of confidence in Scale-down models and CPP ID**

  (**Risk**: Studying too-narrow ranges could mis-classify parameters)

  (**Risk**: Justifications for scale-down model qualification & residual uncertainty did not convince that Design Space limits were valid at scale)

  **Questions remained regarding how change would be managed within the Design Space**

  (**Risk**: Product Quality could drift far from clinical experience without oversight)

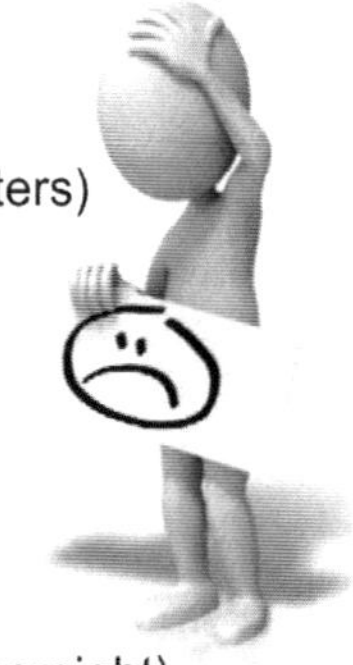

## How Design Space is Defined is Critical

- ICH tells us what a Design Space is…
  - The **multidimensional** combination and interaction of input variables (e.g., material attributes) and **process parameters** that have been demonstrated to **provide assurance of quality**
- …but it does not tell us how to define the Design Space
  - **Should a DS consist of CPPs only, or should noncritical parameters be included? When might the latter be appropriate?**

*"DS should include all relevant parameters required for assurance of product quality…**If you include some control of non-CPPs — or include them somehow into the DS — then data requirements may be lower. If the DS includes CPPs only, then a thorough data package will be needed to convince regulators that you can ignore controls or inclusion of non-CPPs**"*

- From: QbD for Biologics, Learning from the Product Development and Realization (A-MAb) Case Study and the FDA OBP Pilot Program, based on Proceedings of 2010, 23rd CMC Strategy Forum
- by Steve Kozlowski, Wassim Nashabeh, Mark Schenerman, Howard Anderson, Ilse Blumentals, Kowid Ho, Rohin Mahtre, Barbara Rellahan, and Victor Vinci, with Lorna McLeod

## Case Study #2: Drug Substance Site Transfer eCP

Currently Approved GNE/Roche Drug Substance Sites for products A, B, C, D

Future Approved GNE/Roche Drug Substance Sites for products A, B, C, D

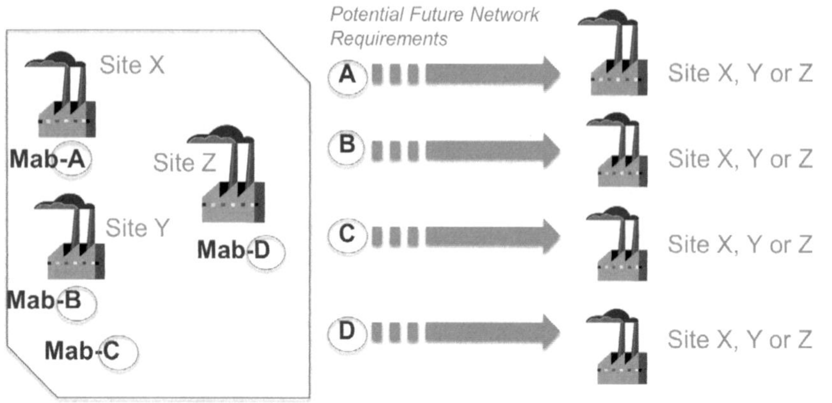


Submit PAS/eCP describing site transfer acceptance criteria for both Site and Product

Execute transfer per defined requirements in eCP

CBE-30 Supplement with data demonstrating acceptance criteria met

11

## Key Questions Regarding Impact of Facility & Process Change Drove Development of Site Transfer Risk Assessment Tools

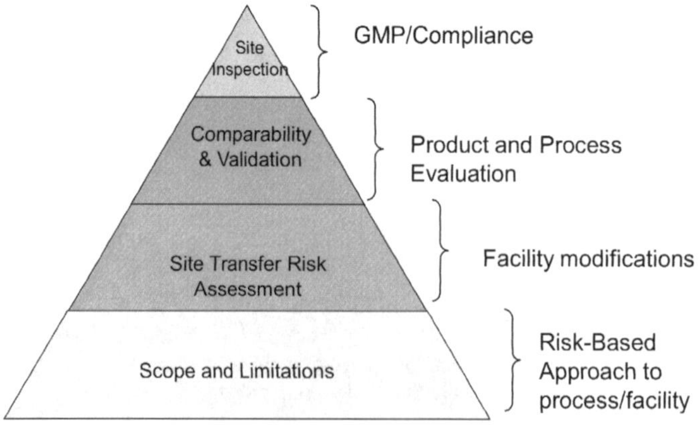


- What products and facilities should be in scope?
- What are the impacts of 'facility fit' changes?
- What should be tested to confirm comparability?
- What should be assessed in considering GMP or inspectional risk?

## Objectives & Outcomes for the eCP

### There was 1 main objective for the eCP filing:

Justify that historical experience with Genentech & Roche's network of production facilities and product characterization was sufficient to support product transfer with a CBE-30 approval.

**Accomplishments:**

- Agreed upon scope: products and facilities for which the Sponsor has sufficient knowledge & experience
- Justified that historical knowledge of product characterization, stability & process performance is sufficient to support comparability with post-approval real-time stability commitment
- Assured that only facilities in "State of Compliance" are allowed, so that PAI may be waived.
- Agreed future products and Facilities could be cross-referenced to the eCP provided they meet the pre-defined criteria

•**Agreement on scope & criteria was reached, and eCP approved ✓**

•**Subsequent transfers meeting eCP criteria approved with CBE-30 ✓**

## Lessons Learned

- Critical to invest in characterization of product quality early, and to confirm Agency agreement with MoA that will drive CQA identification early
- Regulators are open to moving away from traditional approaches to process and product controls
- Degree of "regulatory flexibility" to be expected is directly related to strength of the justification and scope
- Any risk-based decision that removes regulatory pre-approval is going to be highly scrutinized & burden of proof is on manufacturer
- The expectation that changes within the Design Space might be treated as free from regulatory pre-approval (i.e, "not considered a change") is a very high bar for biotech products

## Genentech & Roche's Perspective on what we've gained by implementing and what it cost...

- **Risk assessment tools are valuable to systematically categorize risk in the overall Control Strategy throughout portfolios and across the lifecycle**
- **Cost of RA development is low, the tools and benefits are fully recyclable**
  - Ensure systematic, objective application of historical and SME knowledge
  - Creates a common decision making framework and language to talk to Regulators
- **Significantly enhances assurance of robust product quality**
  - Thorough, integrated evaluation of CQAs during process characterization enhances overall Control Strategy robustness
  - Definition of CPPs is now much more strongly linked to product quality
  - Reduces failure of future process changes and transfer due to incomplete process knowledge
- **Net cost of bioprocess characterization comparable to other commercial Mabs**
- **Streamlining of Control System resulted in some commercial testing savings**
  - Systematically justify controls, avoiding undue or redundant "check-box" testing

## QbD Needs a Makeover!

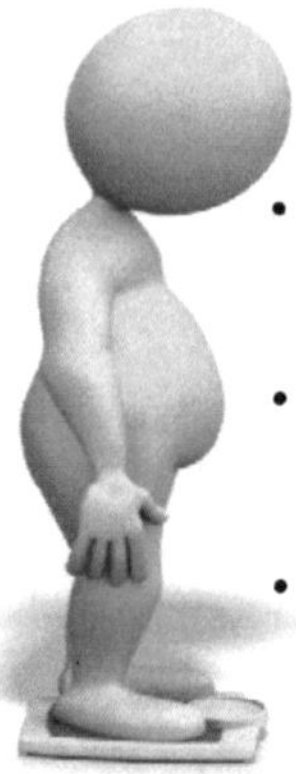

- Need to create a consistent vision of what "QbD" implementation means for all audiences
  - Focus on the value created by robustness of product quality & supply
  - Emphasize QbD as a broad overall paradigm of lifecycle Control Strategy risk management that can in turn justify innovative regulatory pathways, rather than as a means itself to reduced regulatory oversight.
- Need greater clarity about what standards must be met to allow full realization
- Realistic discussion is needed about the balance of risk mitigation -vs- management both industry and regulators can be satisfied with
- Need to improve communication of the overall risk picture in the dossier to facilitate evaluation

## What's Next for Genentech & Roche QbD?

- **Roche believes the benefit of continued implementation of the QbD paradigm is high** We will continue to implement QbD approaches across our biotech portfolio

- **Creating predictability of global regulatory change management is important to us.**
  - We will continue to explore ways to achieve Design Space approval globally
    - Improve process models to the extent practical
    - Improve the clarity of the PALM plan so that measures intended for risk-management of post-approval change are more clearly understood and accepted
    - Explore alternate solutions to Design Space approval that move us forward, but carry less risk. For example...
      - Including all unit operations and non-CPPs in our Design Space definition
      - Limit Design Space to operations whose scale down models are mechanistically understood (i.e., exclude the bioreactor?)
      - Explore reduced, but not absent, pre-approval oversight for changes within the design space?

## Acknowledgments

Dana Andersen
Mary Cromwell
Julia Edwards
Christof Finkler
Christian Hakemeyer
Reed Harris
Brian Kelley
Elisabeth Kirchisner
Andy Kosky
Nathan McKnight
Paul Motchnik
Ron Taticek
Vassia Tegoulia
Felix Kepert
QbD LM Core Team
Perjeta Technical Development Team
GA101 Technical Development Team

Roche

# FDA OBP QbD Pilot Program: Key Outcomes and Learnings

QbD Round Table Workshop
Accelerated Biopharmaceutical Development (AccBio) 2013
Coronado Island, CA
Feb 27, 2013

Ranga Godavarti (Pfizer)
Kris Barnthouse (JnJ)



## Pilot Program Overview

- **Accepted in OBP QbD Pilot April 2009**
- **Monoclonal Antibody manufactured by typical platform process**
- **First meeting: High level overview and approach**
  - General overview of proposed QbD approach
- **Second meeting: CQA determination**
  - Discussion risk assessment approach; structure-function studies; CQA determination
- **Third Meeting: Discussion on design space**
  - Detailed discussion of approaches to parameter criticality determination
  - Definitions of design space
- **Fourth Meeting: Discussion on Control Strategy**
  - Holistic approach to control strategy focusing on raw materials, facility, parametric, in-process, release and stability
- **Program terminated-did not file BLA**

## QbD Meeting 2: Approach to Defining CQAs

- **Initial risk assessment: risk ranking and filtering tool**
  - Discussed overall approach
    - Severity (potential impact on safety and efficacy)
    - Uncertainty: Amount and type of available data (prior knowledge, pre-clinical, clinical etc)
    - Did not consider occurrence (process) or detection (testing)
    - Outcome from initial assessment: Majority of attributes were CQAs due to lack of knowledge
- **Structure-function studies performed on a subset of QAs**
    - Isolated enriched product variants for bioactivity, rat PK studies
    - Characterized product recovered from patient sera samples from clinical PK studies
- **Second risk assessment following results from S/F studies**
    - Resulted in assignment of CQAs and non-CQAs
    - Discussed examples of proposed control strategies for a CQA and a non-CQA



## QA Criticality and Control Strategy Process

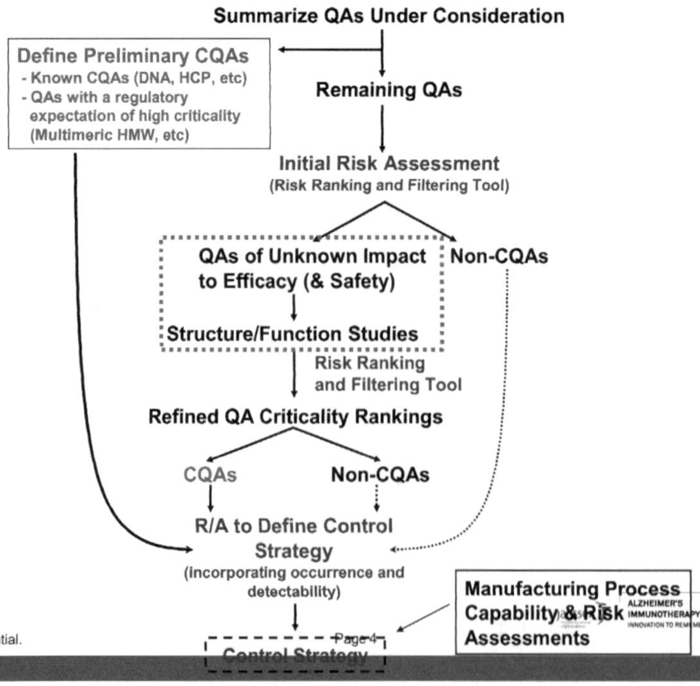

# Key Learnings from CQA F2F Discussion

- General agreement on overall RA approach to assign criticality
  - Final decision on CQA assignment will require full BLA review
- Acceptance criteria for non-CQAs should be in place for comparability
- Structure-function studies
  - General agreement with applicability of S/F data to set CQA ranges. Appreciated the amount of work done to understand QA criticality
  - Concern on MOA understanding and relationship of safety and efficacy to animal models
- Control of non-CQAs
  - Provided example of attribute deemed non-CQA based on S/F studies: Sufficient to monitor attribute only when significant process changes made?
  - Cautious agreement; depends on significance of changes; may require periodic monitoring
  - Concern over non-CQAs-how will they be managed
- Early phase, non-clinical and structure/function data <u>CAN</u> be leveraged to assign CQA ranges outside clinical experience; however phase 1/2 studies can <u>contribute</u> <u>to</u> but are <u>not sufficient</u> basis for acceptable ranges for CQAs
  - Functional studies outlined were considered appropriate; but adequacy of studies based on totality of data available upon BLA review



# QbD Meeting 3: Design Space

- **Described approach to scale-down models**
  - Bioreactor and column chromatography scale-down criteria and data
- **Detailed description of multivariate DOE studies**
  - Risk assessment approach to determining parameters to study
  - Rationale and justification for CPP and non-CPP designation
  - Experimental designs and data
  - Examples of bioreactor and column chromatography design space
- **Proposed example of adaptive control strategy on Mabselect Protein A step**
  - Wash concentration adjusted as resin ages to improve product recovery while meeting LMW target

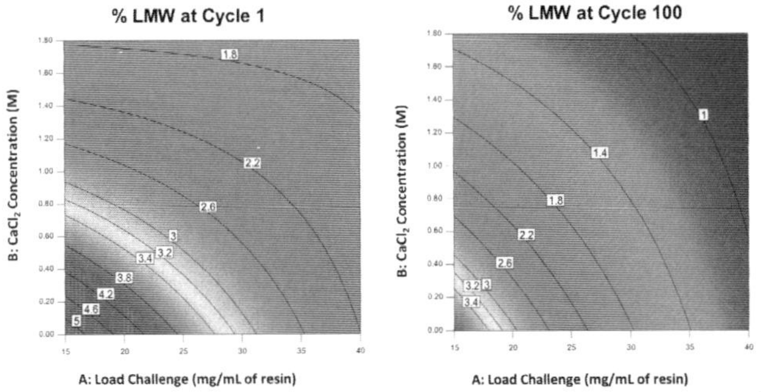

## Key Outcomes and Learnings from Design Space F2F Meeting

- **Strong desire to align to ICH definitions: example 'Critical Process Parameter'**
  - Did not like 'new' terms such as KPPs. Ok to use internally but not file
- **Appreciated amount of data provided conveying process understanding**
  - But unclear on why certain parameters were critical and not others
  - Discussion on ICH Q8 definition of CPP - how does one define '*parameter variability*' and '*impact on CQA*'?
  - Considerable interest at Agency at how will excursions outside of specified ranges of non-CPPs be managed. How will FDA be involved?
    - Agency not comfortable from a regulatory reporting perspective (BLA and lifecycle) as it would provide a lot of latitude for the sponsor
  - Agency comment:
    - As criticality is a continuum, it is not clear to us what is gained by treating parameters deemed 'critical' in a fundamentally different way from those deemed 'non-critical'.
    - Reporting requirements should reflect that continuum of risk for non-CPPs

janssen ALZHEIMER'S IMMUNOTHERAPY PROGRAM INNOVATION TO REMEMBER Pfizer

Confidential. Page 7

## Key Outcomes and Learnings from Design Space F2F Meeting

- **Overall agreement on approaches presented to describe design space**
  - Lot of discussion on relevance of scale-down models
  - Desire to see details of scale-down model qualification in the BLA prior to approving design space
  - Agreed with an approach of adaptive control strategy
- **Opted to not file a design space**
  - Decision based on multiple factors
  - Concern around change management within design space. How will residual risk associated with a design space be dealt with?
  - Agency expectation that changes within design space need to be reported
  - Statistical equivalence of scale down models (instead of lack of difference) and batch number issues
  - Not clear the DS approach would be value added (implementation question)

janssen ALZHEIMER'S IMMUNOTHERAPY PROGRAM INNOVATION TO REMEMBER Pfizer

Confidential. Page 8

## F2F Meeting 4: Control Strategy – key learnings

- Control Strategy elements
  - Discussion on CS design, and elements included.
  - Agreement that the elements should generally include those explicitly defined in ICH Q10/Q11. Also, agreement that CPV is an essential part of CS.
- Design of CS using Risk Assessment Approach
  - Discussion and agreement with using a risk assessment (FMEA-type) to assess and refine the design of the control strategy.
  - Discussion on why an RPN cutoff was not described. The concept was discussed to use the results directionally, i.e. to systematically assess relative risks, but not make a decision to mitigate versus accept residual risks based solely upon the RPN score.
- Specific CQA CS Examples
  - Agency generally agreed with CS examples for CQAs and non-CQAs
  - Open to a PAT approach for one CQA (HMMS): Example of downstream adjustments to remove HMMS based on bioreactor redox potential and filtration/turbidity correlations
  - Agency was generally supportive of the PAT approaches presented. Ultimately, acceptance, however, would be a reviewer decision.
  - Agency also recommended that the sensitivity of analytical test methods to changes in molecular variants be considered

janssen ALZHEIMER'S IMMUNOTHERAPY PROGRAM INNOVATION TO REMEMBER Pfizer

Confidential. Page 9

## F2F Meeting 4: Control Strategy – key learnings

- **Deviation Management**
  - Alliance proposed a tiered strategy for deviation management. Tiers were based on whether the excursion is inside/outside of DS, and/or the degree of available knowledge to assess product quality impact (when inside DS).
  - Discussed whether a deviation beyond the DS would require notification prior to product release.
    - Agency provided feedback that reporting requirements for unintended changes (even one-off deviations) can vary depending upon detectability of the impact and its potential to impact patient safety.
- **Integration into QS – Change Management Strategy**
  - Alliance proposed a tiered strategy for change management similar to the approach described above. Tier drives implementation strategy and reporting classification.
  - Reporting classification will continue to depend upon an assessment of impact on safety and efficacy as per 21 CFR 601.12.
  - Discussed an example where a bioreactor scale-up from 6000-L to 12000-L could be done based on provided process knowledge and be filed as a CBE. General agreement with approach, but it would be a review decision.

janssen ALZHEIMER'S IMMUNOTHERAPY PROGRAM INNOVATION TO REMEMBER Pfizer

Confidential. Page 10

## Outstanding Open Questions for Future

- **Structure Function studies that include detailed non-clinical studies, in-vivo patient sera recovery studies are a big investment.**
  - How can sponsors leverage these data such that the value of these investments are realized in setting an effective control strategy?
- **Where do we see the value of filing design space?**
- **Risk assessment approaches: 'to cut-off or not to cut-off'. How do we get past the criticality continuum loop?**
  - Criticality is a continuum – no 'cut-off' to distinguish critical from non-critical. So how do we distinguish critical from non-critical? What is the 'cut-off'?
  - Use a cut-off RPN: How did you come up with and justify the cut-off? Criticality is a continuum!
- **How do we define or quantify 'parameter variability' and 'CQA impact' in ICH Q8 definition of CPP?**
  - How do we reconcile what parameters get reported in a BLA?
- **What is the future of expanded change protocol?**
  - Post-approval change management
  - Management of regulatory commitments



## Acknowledgements

- QbD Core team from Pfizer and JAI/JnJ
- FDA OBP members

U.S. Food and Drug Administration
Protecting and Promoting Public Health
www.fda.gov

# Unique Challenges of Quality by Design (QbD) for Biotechnology Products

Barbara Rellahan, MS, PhD
Team Leader DMA/OBP/CDER
SBE AccBio/QbD Conference
February 27, 2013

U.S. Food and Drug Administration
Protecting and Promoting Public Health
www.fda.gov

# Disclaimer

Some views expressed in this presentation are my own, and do not necessarily reflect official FDA policy.

## Presentation Outline

1. Critical quality attribute identification challenges
2. Critical process parameter identification challenges
3. Challenges for development/evaluation/communication of a design space and control strategy
4. Opportunities and benefits of QbD for biotechnology products

U.S. Food and Drug Administration
Protecting and Promoting Public Health
www.fda.gov

## OBP QbD Pilot Program

*FR Notice July 2, 2008*

- To define clinically relevant attributes for protein products (regulated by OBP) and link them to manufacturing processes
- To consider quality-by-design (QbD) approaches to unit operations in supplements (10) as well as original applications (5)
- To explore the use of protocols submitted under 21 CFR 314.70(e) and 601.12(e))
- Notice of Extension of Deadlines and Increase in the Number of Original Applications to 8
  - Federal Register/Vol. 74, No. 179 /Sep 17, 2009

# OBP QbD Pilot Update

- Applications Accepted in QbD Pilot
    - 24 Meetings have been held for the six applications

– 6 Original Applications
  - 5 Monoclonal Antibodies and 1 Fc Fusion Protein
  - 1 QbD original submission BLA received in 2012.

– 4 Post-approval Supplements
  - 2 with site transfers to ensure Office of Compliance participation
  - 1 post-approval supplement with eCP approved in 2010 (multi-product, multi-site)



# Structural and Functional Complexity

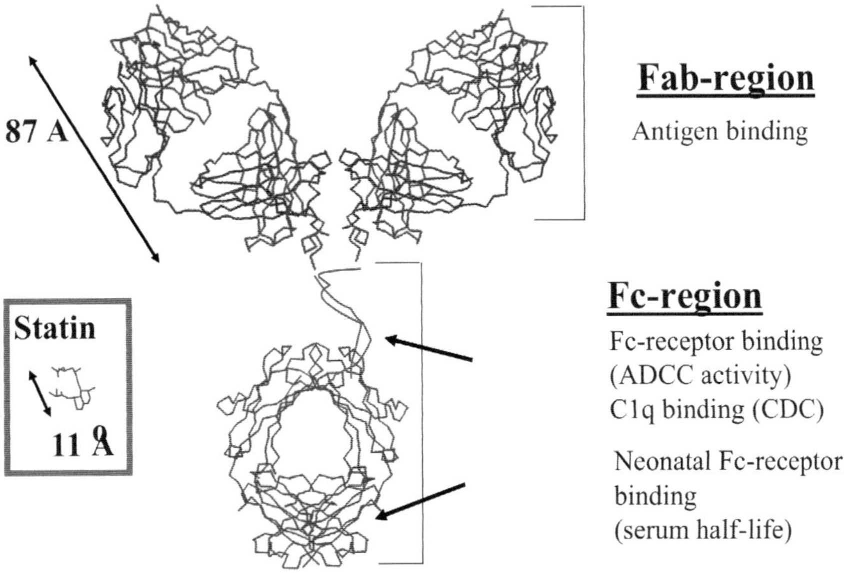


Jefferis et al. *Immunology Letters* (2002) 82; 57-65. & PDB 2IG2, 1FC1

U.S. Food and Drug Administration
Protecting and Promoting Public Health
www.fda.gov

## Post-Translational Heterogeneity

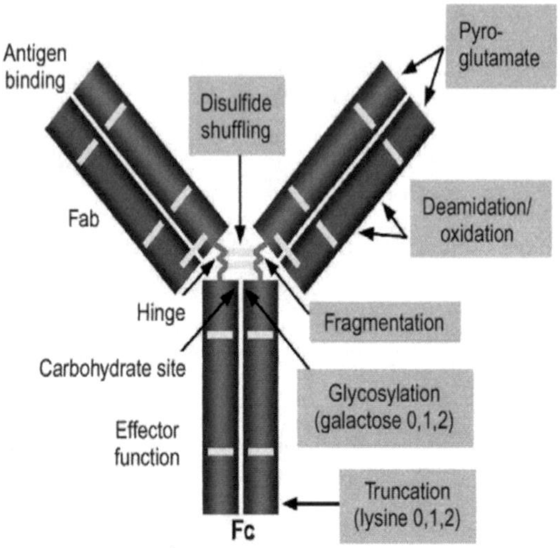


Arthur J Chirino & Anthony Mire-Sluis
*Nature Biotechnology* **22**, 1383 - 1391 (2004)

Aggregation/particulates

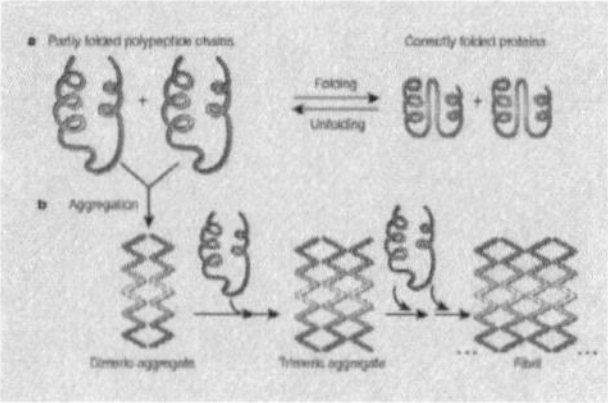

Fragmentation

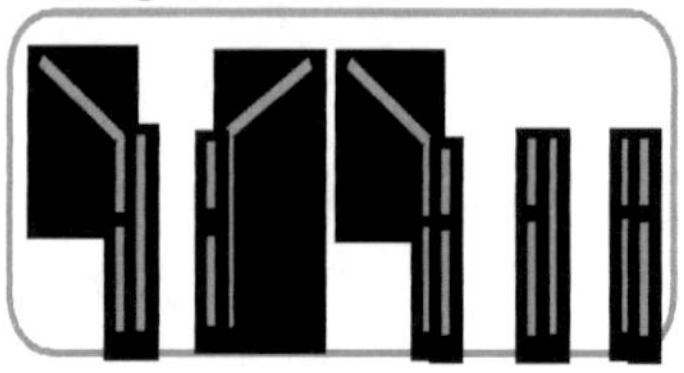

Process related impurities
- Host Cell Proteins
- Host Cell DNA etc.

U.S. Food and Drug Administration
Protecting and Promoting Public Health
www.fda.gov

## Identification of a Product's Critical Quality Attributes (CQAs)

The structural and functional complexity of biotechnology proteins makes identification of a product's critical quality attributes very challenging.

1. Large number of quality attributes need to be assessed for patient impact. Potential for attributes to interact and/or impact stability (e.g., oxidation and free thiols contributing to the formation of aggregates).

2. Many attributes are present at levels near or below the limit of detection. Issues arise on how these attributes should be evaluated and controlled.

U.S. Food and Drug Administration
Protecting and Promoting Public Health
www.fda.gov

## CQA Identification

An extensive dataset on the potential attributes to impact product safety and efficacy must be developed to inform the CQA risk assessment.

- Variants identified by extensive structural characterization analyses/literature information.
- Specific methods developed to detect and monitor attributes present at significant levels and considered higher risk
- Pre-formulation characterization studies to understand the chemical and physical stability of the product under various conditions (e.g., temperature, oxidation, pH, buffer and ionic strength)
- Use small scale development and stress stability studies to provide information on the formation of product variants.
- Variants are enriched or synthetically created and assessed for impact.
- Variants isolated from pre-clinical and/or clinical samples.

**Lack of prior product and platform knowledge presents a real hurdle for implementation of QbD**

U.S. Food and Drug Administration
Protecting and Promoting Public Health
www.fda.gov

## Potential for attributes to interact and/or impact stability

Example: Sponsor found that the level of low MW free thiol species present in the bioreactor correlated with the % of under disulfide bonded (UDB) heavy chain present and the potential for aggregates to form.

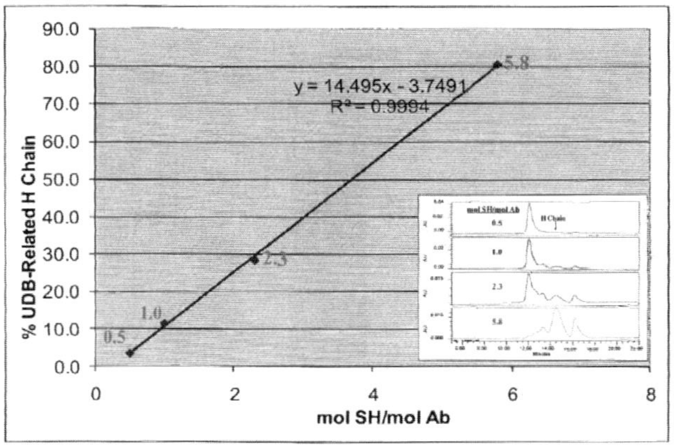


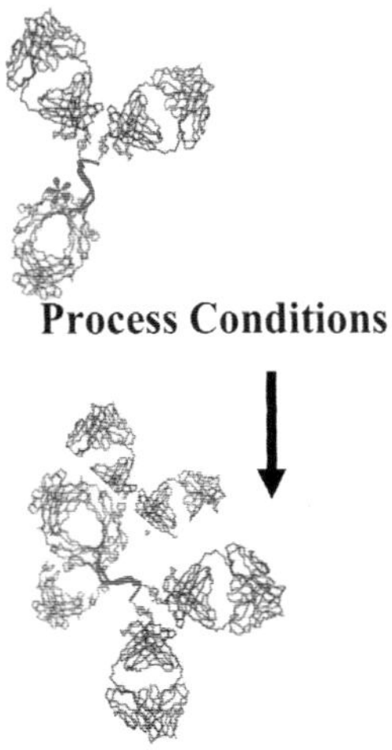

U.S. Food and Drug Administration
Protecting and Promoting Public Health www.fda.gov

## Assessing the Impact of Modifications Such as Deamidation and Oxidation

➢ In many cases these are present at very low, well controlled levels though they may increase during storage.

➢ Can be challenging to isolate sufficient quantities to assess their impact on bioactivity, stability and to determine whether they enhance the formation of additional variants such as aggregates.

➢ Using stressed conditions to increase the level often results in the generation of variants not normally observed so that a direct correlation between oxidation or deamidation at a specific site and product impact can not be made.

➢ Sponsors often have to default to assigning a higher risk level due to uncertainty.

U.S. Food and Drug Administration
Protecting and Promoting Public Health www.fda.gov

## Quality Attributes Present at Low Levels

Question to OBP: In certain instances [the sponsor plans to use] results from 'relevant' stressed conditions to determine the maximum level of the [attribute] which could potentially occur during manufacturing and storage as an input into the Severity Assessment. The stated purpose of this is to avoid making unrealistic worst case assumptions.

Sponsor used two conditions to increase the attributes level and determined whether there was an impact to bioactivity or pK (using a FcRn binding ELISA).

U.S. Food and Drug Administration
Protecting and Promoting Public Health
www.fda.gov

## Quality Attributes Present at Low Levels

It was not clear that this approach would be acceptable since process capability should not be included in the CQA risk assessment (should only be severity x uncertainty). Sponsor was requested to provide more information/data to justify the approach.

However it is acceptable to use stressed samples to generate product samples with increased attribute levels. Attribute levels need to be high enough to ensure adequate characterization of impact.

When a CQA is present at very low, well controlled levels, and was ranked a CQA due to high uncertainty, it may be acceptable to limit the control strategy for this attribute to inclusion in comparability studies

U.S. Food and Drug Administration
Protecting and Promoting Public Health
www.fda.gov

## Quality Attribute Acceptance Criteria

Sponsors are attempting to use their increased quality attribute knowledge and risk assessments to broaden the acceptable range for some attributes so that they extend well outside the range that were observed in the clinical trial (s).

Example:
Clinical range ≤ 1.7%
Trend limit ≤ 3.0% (supported by manufacturing/characterization study ranges)
Comparability limit ≤ 10%

While the ranges requested for the attribute itself may be acceptable, sponsors need to provide process understanding to explain the conditions that would result in an attribute being outside the trend limits, and data to demonstrate that those conditions do not result in a change to other attributes.

U.S. Food and Drug Administration
Protecting and Promoting Public Health
www.fda.gov

## CQA Identification Deficiencies

- Inclusion of process capability or detectability in the CQA risk assessment.
- Absence of complete listing of potential quality attributes in the assessment (e.g., exclusion of secondary mechanism of action from evaluation).
- Poor application of risk assessment tools (e.g., underestimation of the impact on immunogenicity, PK or safety because product contained low levels of the attribute and there was no impact observed in the clinic).
- Lack of assessment of attribute interactions and/or impact of attributes on product stability.

U.S. Food and Drug Administration
Protecting and Promoting Public Health
www.fda.gov

## Critical Process Parameter (CPP) Identification

The large number of complex attributes that need to be evaluated in DoE studies and at full scale pose a challenge in regard to resources and analytical capability.

U.S. Food and Drug Administration
Protecting and Promoting Public Health
www.fda.gov

## CPP Identification Challenges

➢ **For CPP identification, discriminating between a statistically significant change to product quality and one that is 'practically' meaningful.**

For example, a change from 0.3% aggregates to 0.31% aggregates may be statistically significant, but not a practically meaningful change.

One sponsor developed a mathematical formula based on the degree of attribute variation caused by parameter variation observed in their DoE studies compared to the desired acceptance range for the attribute. This provided an objective criteria on which to base their assessment of parameter criticality.

All sponsors will need to specifically explain how this distinction will be made.

U.S. Food and Drug Administration
Protecting and Promoting Public Health
www.fda.gov

## CPP Identification Challenges

**The Goldilocks Paradigm**

➢Determining how much to vary a parameter in the univariate and/or multivariate studies.

**Parameter variability**

| | |
|---|---|
| Too broad | Everything CPP |
| Too narrow | Nothing CPP |
| "Just right" | CPPs correctly identified |

U.S. Food and Drug Administration
Protecting and Promoting Public Health
www.fda.gov

## CPP Identification Challenges

**The Goldilocks Paradigm**
**How do we define "Just Right"?**

Comparison of process ranges studied

| Process parameter | Firm 1 range | Firm 2 range | Firm 3* range |
|---|---|---|---|
| pA elution pH | 3.75-4.10 (0.35) | 2.8-3.1 (0.3) | 3.9-4.3 (0.4) |
| CEX elution pH | 7.7-8.3 (0.5) | 7.9-8.1 (0.2) | 5.0-5.6 (0.6) |
| AEX elution pH | 7.1-8.5 (1.4) | 7.8-8.2 (0.4) | |

U.S. Food and Drug Administration
Protecting and Promoting Public Health
www.fda.gov

## CPP Identification Challenges

- **Uncertainties present in the cell growth process.**

  The cell growth process is extremely complex. It can be challenging to develop representative small scale models and to characterize all sources of variability (e.g., raw material, genetic variability).

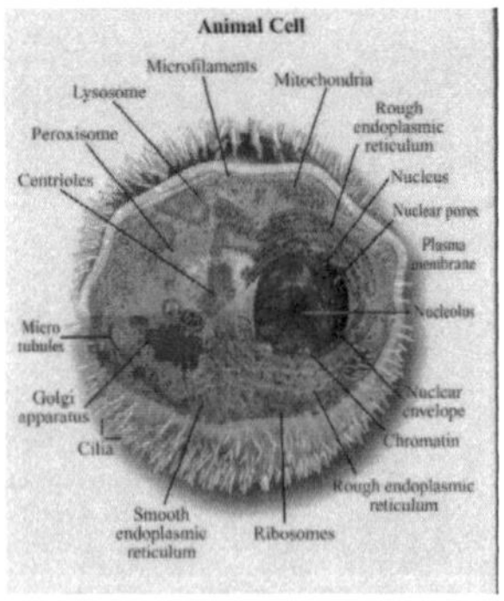


  Cell bank stability

  Early seed train steps are assumed not to impact product quality because no product is accumulated. However, conditions experienced during scale up may impact the biology of the cells which in turn can impact product quality.

U.S. Food and Drug Administration
Protecting and Promoting Public Health
www.fda.gov

## CPP Identification Deficiencies

- Deficiencies in CQA identification process.
- Insufficient information on the risk assessments performed (and their results) to justify parameter inclusion/exclusion from the DoE studies.
- Ranges used were not always broad enough to assure they would identify CQA impact.
- Small scale studies not demonstrated to be representative of full scale process.

U.S. Food and Drug Administration
Protecting and Promoting Public Health
www.fda.gov

## Design Space Challenges

**Design space model fit/adequacy.**
Analysis of variance to ensure that the model fit is significant and to determine $R^2$ values. $R^2$ value threshold of $\geq 0.5$ is used to determine if model is adequate for process modeling.

In cases where $R^2 < 0.5$, many sponsors have gone back and examined slightly different parameter ranges to increase confidence of the model.

Many sponsors also tested the 'predictiveness' of their models using small scale models to determine predicted and actual results.

U.S. Food and Drug Administration
Protecting and Promoting Public Health
www.fda.gov

## Design Space Challenges

**Industry (and Agency)**

➢Communicating (or reviewing) the overall control strategy, the data that supports that strategy and the planned lifecycle management strategy to the regulatory authority.

**Agency**

➢Evaluating whether the proposed design space provides sufficient assurance that product quality will be maintained.

➢How much regulatory flexibility does the design space support (movement within DS, acceptable range for non-CPPs etc).

U.S. Food and Drug Administration
Protecting and Promoting Public Health
www.fda.gov

## ICH Q11 Example 5a. Possible Control Strategy Summary – Biotechnological Products

| Drug Substance CQA | Control Strategy for drug substance CQA | Section(s) in CTD where detailed information is located |
|---|---|---|
| Contaminant (Viral Safety) | Summaries of viral safety information for biologically-sourced materials | 3.2.S.2.3 |
| | Detailed information including for materials of biological origin, testing at appropriate stages of production and viral clearance studies | 3.2.A.2 |
| Residual Host Cell Proteins | Design Space for an individual unit operation (e.g. see Example 3) | 3.2.S.2.2 |
| | Target range for consistent removal assured by validation | 3.2.S.2.5 |
| | Analytical procedures and their validation | 3.2.S.4.2 , 3.2.S.4.3 |
| Specific Glycoforms | Controls implicit in the design of the manufacturing process including a summary of process control steps (e.g. cell culture conditions, downstream purification, holding conditions etc.) | 3.2.S.2.2 |
| | Characterisation to justify classification as CQA (cross reference to non-clinical/clinical sections if relevant) | 3.2.S.3.1 |
| | Control of Critical Steps, Testing program and specifications | 3.2.S.2.4, 3.2.S.4.1 |
| | Justification of specification | 3.2.S.4.5 |
| | Stability | 3.2.S.7 |

U.S. Food and Drug Administration
Protecting and Promoting Public Health
www.fda.gov

## Communicating Knowledge and Control Strategies

**Compiled View of Product Controls for Select Quality Attributes**

| Quality Attributes | Drug Substance | | | | | | | | Drug Product | | | | | | | |
|---|---|---|---|---|---|---|---|---|---|---|---|---|---|---|---|---|
| Element: | 1 | 2 | 3 | 4 | 5 | 6 | 7 | 8 | 1 | 2 | 3 | 4 | 5 | 6 | 7 | 8 |
| | | | | | | | | | | | | | | | | |
| | | | | | | | | | | | | | | | | |
| | | | | | | | | | | | | | | | | |
| | | | | | | | | | | | | | | | | |
| | | | | | | | | | | | | | | | | |

| Color Coding |
|---|
| Critical quality attribute |
| Non-critical quality attribute |
| Parameters within element are monitored |
| Parameters within element are controlled |
| Element is neither monitored or controlled |

| | Elements | |
|---|---|---|
| 1 | | |
| 2 | e.g., in process testing, release/stability testing, raw material testing etc. | y |
| 3 | | |
| 4 | | |
| 5 | | |
| 6 | | |
| 7 | | |
| 8 | | |

U.S. Food and Drug Administration
Protecting and Promoting Public Health
www.fda.gov

## Design Space Challenges
## Communicating Knowledge and Control Strategies

**Examples of information to submit (list is not complete)**

- Provide detailed explanation of all risk assessment tools used. Include justification for scoring system and cut-offs used.
- For CQA identification process provide a detailed summary narrative of the assessment process for each quality attribute. Include summary data from characterization studies performed to assess attribute criticality and links to literature cited.
- For CPP identification process provide information on statistical analysis performed, summary narratives for all unit operation steps, full report for 2-3 unit operations (e.g., bioreactor, steps were few CPPs ID'ed).
- Detailed explanation of how parameters not included in the design space will be managed.

## Opportunities/Benefits

➢ The pilot program applications have demonstrated that pursuing QbD for biotechnology products results in a significant increase in a sponsor's understanding of how product attributes impact the product's safety and activity, and of how the process impacts product attributes.

➢ Many sponsors are attempting to use this increased knowledge to optimize their processes and incorporate more flexibility into their operating parameters, and develop a focused control strategy.

U.S. Food and Drug Administration
Protecting and Promoting Public Health
www.fda.gov

## Acknowledgements

Patrick Swann, PhD
DMA/OBP/CDER

DMA/OBP Regulatory Review Branch

**A Robust and Disposable Flow Through Purification Train For Post-ProteinA Mab Processing**

Bam , N., GlaxoSmithKline
Stone, M., EMD Millipore
Cataldo, B., EMD Millipore
Galipeau, K., EMD Millipore
Phillips, M., EMD Millipore
Kozlov, M., EMD Millipore
Originally presented on: 2/24/2013 17:55 - 18:05

In this work, we propose a novel mAb purification process replacing typical post-proteinA steps of bind/elute CEX chromatography, flow-through AEX chromatography and virus filtration with a fully-connected flow-through purification (FTP) train. The FTP train — comprising of 3 different single-use adsorbers, in-line pH adjustment, and virus filtration — was specifically designed to remove process- and product-related impurities like host-cell proteins, DNA, leached-PrA, virus, and aggregates in flow-through mode thereby eliminating the need for hold tanks and bind/elute CEX step. Experimental data highlighting robust performance (yield and impurity removal) of the fully-connected FTP train with multiple mAbs will be presented and discussed in detail. This novel purification process has the potential to not only improve manufacturing flexibility, productivity and economics but also enable fast and simple process development without compromising process robustness.

**Accomodating Filter Variability in Process Design**

Lutz, H., Millipore Corporation

Originally presented on: 2/24/2013 18:05 - 18:15

Lot variability in raw materials and differences in performance at different scales of operation may impact manufacturing scale operation. This talk will characterize and identify the magnitude and sources of process variability and scaling differences for a variety of filtration unit operations using performance models and test data. This variability can be accomodated through operating practices and the use of both a scaling factor and a safety factor in sizing. Values for these factors will be derived for a variety of filters and the implications for process risk identified.

**Bayesian Approach to Design Space for Drug Product Dilution and Fill**

Boulanger, B., Arlenda
Lebrun, P., Arlenda
Scherder, T., Arlenda
Originally presented on: 2/24/2013 18:15 - 18:25

Before proceeding to the filling of syringes, the bulk material must usually be diluted to achieve a desired concentration. Also release of the final product requires that the average content or result is greater or equal to a lower specification limit (LSL). Given the inherent uncertainty for the estimate of the content of the bulk material, there is a risk that the dilution factor derived from a point estimate can lead to under dosing the final product. To avoid this possibility, a typical practice is to apply a margin of safety by slightly under-diluting the sample to ensure specifications will be met. But this carries the risk of over-dose of the final product. In addition, the final release of the product is evaluated using the same assay, ie the release also is made with an uncertainty. Finally, the precision of the assay is itself determined with some uncertainty. In this poster we'll present a Bayesian model and simulation technique that take into account the three uncertainties to optimally dilute the bulk: the estimate of content of the bulk to derive the dilution factor, the estimate of the content of the final product and the estimate of the precision of the assay (repeatability and intermediate precision). The results will be presented to minimize the risk of rejecting a lot while minimizing the risk of over dosing the final product. In addition, various sample size [number of runs and number of replicates per run] will be examined to propose a practical sample size given time constraints within the laboratories.

**Benchmarking Study to Evaluate Trends for Viral Barrier Implementation in Commercial Cell Culture Processes**

Delfosse, S., Keck Graduate Institute
Sathavipat, M., Keck Graduate Institute
Hsu, N., Keck Graduate Institute
Croughan, M., Keck Graduate Institute
Originally presented on: 2/24/2013 18:25 - 18:35

Nearly twenty years have passed since Genentech reported the first large-scale contamination of commercial cell culture operations by Mouse Minute Virus (MMV). Since then, a handful of large-scale contaminations have been reported along with various strategies, technologies, and best practices for mitigation of adventitious viral and mycoplasma contaminations in upstream cell culture processes. Despite this shared information, many uncertainties exist regarding implementation of viral barriers for upstream cell culture processes. Focusing primarily on High Temperature, Short Time (HTST) treatment of cell culture media, we conducted a benchmark study of the current state of implementation of in-process, upstream viral barriers, and also evaluated the challenges companies have faced in regards to this topic. The results generated are beneficial as a guide for viral risk mitigation efforts, at both established and start-up biopharmaceutical companies. This study surveyed nine leading biopharmaceutical companies, as well as six industry-leading consultants, who were able to share insights into the regulatory landscape and potential hurdles, scalability challenges, best practices for viral inactivation and validation studies, as well as the impact to cost of goods.

**Controlling High Mannose Glycan Level and Optimizing Titer Through a Balanced Modulation of Cell Culture Process and Medium Changes**

Le, N., Amgen

Originally presented on: 2/24/2013 18:35 - 18:45

High mannose glycoform may be an important quality attribute that can affect the efficacy of therapeutic monoclonal antibodies. Meeting comparable high mannose glycan level (%HM) is often important when developing a commercial process to replace the early phase clinical process. Strategies were developed to tune the %HM in order to meet the comparability range while at the same time increasing titer for the commercial process. The strategies utilized a combination of process optimization and medium modification in order to produce the desirable %HM profile. In several process instances, titer had to be sacrificed in order to meet the desirable %HM level; however, this was overcome by modulation of certain medium components. Yet, precise balance between process and medium needed to be made in order to achieve the most desirable state for both titer and %HM. Overall strategies and learning on taking this dual approach to controlling the %HM while improving titer will be discussed.

**Genome-scale metabolic reconstructions of Escherichia coli highlight strain-specific adaptations to environmental niches**

Monk, J., University of California, San Diego

Originally presented on: 2/24/2013 18:45 - 18:55

The number of complete genome sequences for E. coli strains continues to accumulate rapidly. We used 47 of these sequences and 152 draft sequences to generate strain-specific metabolic models using homology mapping of the content from the most recent metabolic model for E. coli K-12 MG1655 (iJO1366) as well as metabolic models of closely related species (Salmonella typhimurium LT2 (STMv1.0), Klebsiella pneumoniae MGH78578 (iYL1228) and Yersinia pestis CO95 (iPC815)). These new strain-specific models predict differences in strain phenotypes and allow for the generation of testable hypotheses for growth substrates and strain-specific auxotrophies. The models can further be used to improve genome annotations, transplant metabolic pathways of interest and predict gene essentiality in pathogenic strains to fight infection.

**Human and Cultural Factors in QbD - Risks and Opportunities**

Scherder, T., Arlenda

Originally presented on: 2/24/2013 18:55 - 19:05

The principles of Quality by Design, including early definition of the factors that affect quality, and implementation of a quality management system, are well recognized as essential to predictability and profitability of any manufacturing process. Yet, are there elements of quality doctrine that have not been exploited to achieve quality and operational goals?

Much focus in QbD programs is given to the appropriate use of scientific methods to establish relationships between process factors and critical quality attributes, aligned with a system to monitor and manage these relationships. These activities, often pursued during development phases, are critical to ensure quality product. However, beyond this focus and timeframe lies substantial opportunity to maintain product quality and profitability, specifically through the recognition and management of longer term human interaction and cultural factors. And, the more abstract the process factors, the greater the need for proactive management of these human factors.

Biologics manufacturing presents a unique combination of challenges to maintaining control and reducing variability, including: production mechanisms that are complex, and often intangible, plus notoriously high measurement variability (both in-process and final product attributes). Hence, it is crucial that variability due to human interaction be continually understood and minimized, throughout the product lifecycle. A culture of systems thinking and continuous improvement is also demanded in this environment. Only then will changes in the process be promptly identified, and new knowledge shared between the functional (and often siloed) groups of early and late stage development, analytical, quality, commercialization, and operations.

The author describes the essential supplement of the lean manufacturing principles of standardized work, systems thinking, and continuous improvement to any QbD program to achieve operational goals. She provides examples in biologics manufacturing for the potential consequences when these principles are overlooked, and the benefits when they are deliberately managed.

**Lessons Learned from a Scale-down Model Qualification Case Study**

Meier, A., Genentech

Originally presented on: 2/24/2013 19:05 - 19:15

Cell culture process characterization studies for a protein production process were performed using a scale-down model. A Quality by Design (QbD) approach was taken in order to improve overall process understanding, with the final outcome being both univariate and multivariate acceptable ranges for each parameter tested. However, applicability of these study results to the manufacturing scale depends on the validity of the scale-down models that have been used. In this particularly complex case study, both small and pilot scale models were qualified. Evaluation of some product quality attributes required further downstream processing through scale-down purification models. As a result, qualification of the cell culture model ultimately relies on qualification with the purification scale-down model as well. The way the models were applied depends both on the product quality attribute being evaluated and the unit-operation being characterized. Our scale-down model qualification approach will be presented and challenges related to applying scale-down results to manufacturing scale will be discussed.

Key Message Bullet Points:

- A QbD approach to cell culture process characterization relies on the use of scale-down models
- Applicability of study results to manufacturing scale depends on the validity of the scale-down models that are used
- Our scale-down model qualification approach will be presented and challenges related to applying scale-down results to manufacturing scale will be discussed

**Manufacturing Incidents Investigation Involving Raw Materials Used For Biopharmaceutical Products**

Wen, Z., Amgen Inc.

Originally presented on: 2/24/2013 19:15 - 19:25

Raw materials used in biopharmaceutical manufacturing processes are critical to the quality of biopharmaceutical products. These materials include media components for cell culture, column resin for purification, excipients and containers for final drug product. The quality of final drug products depends on the quality of all the raw materials used during the manufacturing processes. Raw materials for biopharmaceutical manufacturing are not regulated to the same degree as drug substance and hence impurities, contamination, deficiency of critical components in raw materials can impair the product quality of biopharmaceuticals. For instance, contamination of heparin, glass delamination of vials, and tungsten residue in glass syringes can critically impact drug products. In this presentation, we report a number of incident investigations of lot failures during biopharmaceutical manufacturing that were caused by raw materials. They include black particles found in prefilled glass syringes, tungsten residues that impacted product stability, and plastic particle residues found in prefilled plastic syringes.

**Monitoring of Multiple Product Quality Attributes of Biologics using Stable Isotope Labeled Internal Standard and Mass Spectrometry**

Wypych, J., Amgen

Originally presented on: 2/24/2013 19:25 - 19:35

With the implementation of the quality by design (QbD) concept to drug development, monitoring product quality attributes (PQAs) of biologics during process and product development has become increasingly critical. Mass spectrometry has been widely used for quantification of modifications in therapeutic proteins. In addition the use of internal standards provide unique benefits for accurate and precise quantification, since all variability in sample handling is minimized as the internal standard is introduced into the sample for analysis prior to any sample handling steps, such as proteolytic digestion and chromatographic separation, etc. Here, we describe the introduction by in vivo metabolic labeling of a stable isotope to produce a stable isotope labeled internal standard (SILIS) protein with 15N enriched media. The SILIS protein is the foundation for a novel strategy for automatically monitoring multiple PQAs in a single LC/MS analysis. The 15N labeled biologics was used as an internal standard to monitor multiple PQAs of in-process samples. Precise and accurate quantitation was achieved on a Thermo Scientific LTQ Orbitrap Velos high-resolution mass pectrometer. MassAnalyzer, an in-house developed software at Amgen, was used for automated identification and quantitation of multiple modifications identified in a single peptide map as part of process and product development.

**QbD Strategies for Managing Chromatography Resin Lot-to-Lot Variability Based On Design of Experiments and Monte Carlo Simulation**

Ahnfelt, M., GE Healthcare Life Sciences
Lacki, K. M., GE Healthcare Life Sciences R&D
Bergander , T., GE
Malmquist, G., GE
Originally presented on: 2/24/2013 19:35 - 19:45

Quality by Design highlights the need for better process understanding, both regarding Critical Process Parameters and Critical Raw Material Attributes (CMA's). Chromatographic media contain several potential CMA's, and GE Healthcare is considering a multifaceted raw material framework that focuses on identification of CMA's through dedicated studies and intensified supplier – end user collaboration efforts. The framework will be illustrated by examples describing process characterization studies performed using high-throughput process development tools and chromatography medium batches with varying attributes within respective specification ranges. The underlying principle behind this approach should be contrasted to typical resin screening where different resins are compared, as in the proposed framework all attributes reflect the normal manufacturing variation within one resin product. Special emphasis will be placed on ligand density, one of the top CMA candidates as it can affect chromatographic selectivity. The framework employs HTPD data, Design of Experiments and Monte Carlo simulation to establish the effect of chromatography medium attributes and selected process parameters on CQA's. The evaluation includes process parameter adaptation by looking for parameter interactions that allow an adaptive control strategy where the effect of variations in chromatography medium attributes can be counteracted by changing process parameters. The proposed approach outperforms conventional testing of a few resin lots from normal production by explicitly investigating a potential CMA. When established, the adaptive process control strategy maintains the process within the defined operating window, yet minimizes effects related to chromatography medium lot to lot variability and/or eventually leading to a non-critical assessment of the medium from the raw material classification perspective. The proposed adaptive control strategy will result in a more robust purification step and increased security of supply, and in turn better process economy.

**Single Use Centrifuge Calrification Of High Density CHO**

Richardson, D., Drexel University

Originally presented on: 2/24/2013 19:45 - 19:55

The large scale economic collection of proteins is an important problem for the Biotechnology industry. Generally, proteins are produced by an engineered cell culture where a recombinant plasmid containing the gene for that protein is inserted. Since the cell lines used are living organisms, they must be fed with a complex growth medium. Separation of the desired protein from the fermentation broth poses challenges. Advances in cell culture techniques have greatly increased the cell populations and titers of the target proteins in culture fluid. This increase in upstream efficiency has lead to a bottleneck in downstream processing at the cell-harvest stage. Cell harvesting or clarification of the cell culture fluid is an important process. When the product is extracellular, clarification is used to separate the product from the cells and cellular debris. When the finished product is the cells or intercellular proteins/virus cell harvesting is performed. Most Biopharm companies today are seeking ways to integrate this single use technology (clarification/harvest) into the single use process path. The UniFuge single use centrifuge is fits very well into the single use initiative. This presentation will compare the UniFuge to several cell harvest/clarification processes.

**Use of Annotated Chinese Hamster Reference Genome and Transcriptome to Investigate Regulatory Mechanisms and Mutation Rates**

Jacob, N. M., University of Minnesota
Yusufi, F., Bioprocessing Technology Institute A-STAR
Johnson, K. C., University of Minnesota
Yongky, A., University of Minnesota
Le, H., University of Minnesota
Lee, D., Bioprocessing Technology Institute A-STAR
Karypis, G., University of Minnesota
Yap, M. G. S., Agency for Science and Technology Research (A*STAR)
Hu, W. S., University of Minnesota
Originally presented on: 2/24/2013 19:55 - 20:05

We have assembled the Chinese hamster genome as well as comprehensive transcriptomes of hamster tissues and CHO cells. The diploid genome will serve as a reference for the genomic study of CHO cells, which are known to be aneuploid and have a propensity for chromosomal rearrangement. The annotation of the hamster genome was facilitated by our EST repertoire, which was collected from transcriptionally active liver, brain and spleen tissues as well as a wide range of CHO cell lines under different culture conditions to enhance transcript diversity. Over 95% of ESTs aligned to the assembled Chinese hamster genome, indicating excellent coverage of gene-rich regions. Ab initio gene prediction was performed and regions with homology to mouse, rat and human protein coding sequences were identified. Genome annotation facilitated investigation of synteny with mouse to estimate the degree to which gene order is maintained; mouse orthologs were identified for 85% of the hamster genes.

# AUTHOR INDEX

**AIChE**
120 Wall Street, FL 23
New York, NY 10005-4020

ISBN 978-1-5108-1598-8
9 781510 815988

CURRAN ASSOCIATES INC.
proceedings.com